Pioneer Steelmaker in the West

Pioneer Steelmaker in the West

The Colorado Fuel and Iron Company 1872-1903

by H. Lee Scamehorn

PRUETT PUBLISHING COMPANY
Boulder, Colorado

First Edition
1 2 3 4 5 6 7 8 9 0

Library of Congress Cataloging in Publication Data

Scamehorn, Howard Lee, 1926-
Pioneer steelmaker in the West.

Includes bibliographical references and index.
1. Colorado Fuel and Iron Company. 2. Steel industry and trade — Colorado — History. 3. Colorado — Industries — History. I. Title.
HD9519.C58S28 338.7'66'914209788 76-25462
ISBN 0-87108-099-0

Printed in the United States of America

Preface

Herbert N. Casson, popular historian of the American steel industry, remarked in 1907 that it seemed strange that no one had written the history of the Colorado Fuel and Iron Company, "the wonderful story of Pueblo and the men who made it the Pittsburgh of the West." Indeed, it is difficult to explain why historians for a period of a hundred years have ignored the one enterprise that played a key role in shaping the history of Colorado and the Rocky Mountain region.

This volume, *Pioneer Steelmaker in the West,* endeavors to place in historical perspective the origin and development of the Colorado Fuel and Iron Company and its predecessors during the years from 1872 to 1903. The period has a certain logic of its own. The year 1872 marked the appearance of the Central Colorado Improvement Company for the purpose of assisting the Denver and Rio Grande Railway in building a branch from Pueblo to Canon City. That enterprise was consolidated in 1880 with two other of General William Jackson Palmer's nonrailroad business ventures as the Colorado Coal and Iron Company to erect and operate an integrated iron and steel works, the first of its kind west of the Mississippi River, at Pueblo. Twelve years later, John C. Osgood, president of the Colorado Fuel Company, negotiated a merger of his firm and the Colorado Coal and Iron Company, from which came the Colorado Fuel and Iron Company. He was the chief executive officer of the latter corporation until 1903 when, in order to avoid its bankruptcy, he surrendered control to George Jay Gould and John D. Rockefeller, Jr. That year is the terminal date for this study. With Osgood's resignation, the firm entered a new phase of development characterized by a struggle to transform a regional producer into a manufacturer of national stature.

The emphasis of this study of the Colorado Fuel and Iron Company and its predecessors is on business enterprises and the people who guided the evolution of the firm's fuel and metallurgical operations in Colorado and surrounding states and territories. Despite the fact that these activities date from 1872, in the case of coal mining, they are generally unknown, and the corporation's contributions to the development of the state and region are unappreciated. Historians have, with few exceptions, limited their investigations of the Pueblo enterprise to labor-management disputes while ignoring constructive aspects of the corporation's activities.

This volume stresses the role of business entities and the men who ran them in developing and sustaining heavy industry in Colorado. Labor-management problems, while by no means ignored, have been relegated to positions of secondary importance.

The history of the Colorado Fuel and Iron Company is treated topically and chronologically. The first chapter outlines the evolution of the American iron and steel industry in the nineteenth century, and identifies in Colorado conditions which prompted men associated with the Denver and Rio Grande Railway to establish facilities for metallurgical production in the West. Separate chapters are devoted to the Central Colorado Improvement Company, the Southern Colorado Coal and Town Company, the Pueblo and Denver rolling mills, and the Colorado Fuel Company. The Colorado Coal and Iron Company is credited with two distinct periods of growth. One from 1880 to 1884 was presided over by General William J. Palmer. After he resigned in the latter year, the firm was controlled by Pennsylvania and New York capitalists who could not cope with mounting competition and declining prices. The enterprise's weakened financial condition enabled Osgood to negotiate the consolidation from which emerged in 1892 the Colorado Fuel and Iron Company.

Six chapters delineate the activities of the fuel and iron enterprise during the eleven years it remained under the direction of the "Iowa Group," men associated with Osgood in the management of a vast industrial empire in the Rocky Mountain region. The corporation, like its predecessors, engaged principally in the fuel trade. A plan to reconstruct the steel works was blocked by the Panic of 1893. When the Pueblo plant was finally modernized and expanded at the turn of the century, the cost far exceeded the company's resources; the result it faced was the likelihood of insolvency. Osgood, after having successfully resisted a planned take-over by Chicago financier John W. Gates, voluntarily surrendered control to Gould and Rockefeller.

An epilogue outlines the evolution of the Colorado Fuel and Iron Company and successor corporations from 1903 to the present. This final chapter attempts to place the events of 1872-1903, the focus of this volume, in the context of the Pueblo firm's twentieth-century development. In response to changing marketing conditions, the enterprise gradually reduced its reliance on fuel and emphasized metallurgical production. Economic stagnation led to bankruptcy in the Great Depression, from which emerged the Colorado Fuel and Iron Corporation. An acquisition program inaugurated in 1937 was carried forward over a period of 15 years, enabling the Colorado firm to expand beyond its regional market to acquire national recognition. The altered character of business activites prompted in 1966 a change in name to CF&I Steel Corporation. Three years later, the western enterprise was acquired by the Crane Company of New York, becoming an important element of that conglomerate's international operations.

This study was undertaken at the request of officials of the Colorado Fuel and Iron Corporation. The company, through a gift to the Alumni Development Foundation of the University of Colorado, underwrote expenses involved in conducting research in Denver, Pueblo, Santa Fe, Washington, D.C., and numerous communities of southern and western Colorado where coal and iron mining were once thriving industries. The firm also promised the author full access to all pertinent records in its archives, in addition to freedom in assessing their meaning and in writing a narrative history for the years from 1872 to 1903. These pledges have been fully honored in every respect.

The author has endeavored to let the record speak for itself. The evolution of the Colorado Fuel and Iron Company and its predecessors prior to 1903 has been reconstructed from contemporary data — for the most part, the business records of the enterprises themselves. No attempt has been made to deliberately praise or condemn either business entities or individuals. The manner in which corporations behaved seemed, to officials in decision-making roles at the time, correct; for that reason, it would be improper to evaluate their actions by current standards of behavior. In order to properly understand how and why heavy industry grew in Colorado in the late nineteenth century, the Pueblo firm's actions must be viewed in light of attitudes and values of the time rather than by a code of conduct applicable to an era other than the Gilded Age.

Numerous debts of gratitude were incurred in the preparation of this volume. The author thanks employees and former employees of the CF&I Steel Corporation for generous assistance and cooperation during the research and writing of this study. Special appreciation is expressed to Ralph D. Adkins, C. C. Blalock, Dr. David W. Boyer, Otto Eastlond, Francois Genty, Cameron Kirk, U. E. McLain, Algird C. Pocius, Robert R. Williams, and Richard E. Yates.

Several persons who had been affiliated in one way or another over a period of years with the Colorado Fuel and Iron Company gave generously of their time to acquaint the author with the many facets of the firm's operations. Special thanks go to R. J. Alexander, George A. Beatty, Robert Jackson, Victor Johnson, Roy Reed, and Leslie Van Arsdale of Pueblo; Elizabeth Griffiths and Isaac D. Williams of Canon City; William McShane of Florence; and James H. P. Fisk of Walsenburg.

The professional staffs of numerous libraries, archives, and manuscript depositories contributed in important ways to the research for this study of Colorado's pioneer heavy industry. The author wishes to thank for many courtesies the Western History Collections, Norlin Library, University of Colorado; the Library of The State Historical Society of Colorado; the Western History Departments of the Denver Public Library and McClelland Library, Pueblo; The Charles Leaming Tutt Library, Colorado College; The Western History Department, University of Wyoming; The Kansas State Historical Society; The New Mexico Records Center and Archives; The New Mexico Historical Society; The Colorado State Archives; and The National Archives, Washington, D.C., and the Federal Records Center and Archives, Region 8, NARS, Denver. Delbert A. Bishop and Robert Svenningsen of the last-named depository were responsible for making known to the author the value of federal court records as a source for investigating the history of the American West.

Two colleagues, Robert G. Athearn and Brit A. Storey, critically reviewed the manuscript of this study prior to publication. Their constructive evaluations of content, organization, and style greatly enhanced the historical accuracy and value of the final product. Remaining errors are, of course, solely the responsibility of the author.

Special debts must also be acknowledged to Dianne Anderson for typing the initial draft of the manuscript, and to Marilou, without whose encouragement and support this book could not have been written.

Lee Scamehorn
University of Colorado
Boulder Campus

Contents

1
Introduction

The iron industry was an important segment of the American economy long before the thirteen colonies of the Atlantic seaboard won their independence from Great Britain. Production centered in Pennsylvania but was by no means limited to that area, since the requisite resources — ferrous ore, charcoal, and limestone — abounded in many parts of the country.[1] By the close of the colonial era, primary and secondary manufactures were as a rule separate pursuits. It was more economical to transport metal than ore; for that reason, foundries and mills were usually strategically located in the midst of local markets. Pig and bar iron was shipped from furnaces and bloomeries in remote regions to fabricators of semifinished and finished products in or near population centers.

After the American Revolution, iron production spread beyond the Allegheny Mountains. Pittsburgh and Cincinnati emerged as major foundry and mill towns, obtaining metal from furnaces located near sources of ore and fuel in rural Pennsylvania and Ohio.[2] In the East as well as the West, the output of iron advanced in response to a growing market for farm implements, household wares, machinery for factories, and other goods that reflected the quickening tempo of American economic life.

Railroads became in time an important consumer of iron. They required large quantities of metal for tracks, accessories, locomotives, and rolling stock. Prior to the Civil War, the new mode of transportation was an important impetus for the expansion of furnaces and rolling mills in the United States.[3]

Expanding markets prompted changes of far-reaching consequences in the iron industry. The capacity and efficiency of furnaces increased appreciably, and rolled products overshadowed cast goods. The blowing engine and hot blast, in conjunction with the substitution of mineral fuels for charcoal, lowered smelting costs.[4] The decreasing cost of transportation and the adoption of coke hastened the transfer of primary production from rural settings to urban centers. This relocation marked the beginning of integration in metallurgical manufacturing.[5]

The pre-Civil War industry also experienced a subtle but important change in the nature of its product. Until the mid-century, cast iron was almost universally employed in the fabrication of finished goods; after that date, wrought iron gained ascendancy because its resiliency and ability to withstand pressure made it the ideal metal for the manufacture of rails and structural forms for the construction trade.[6]

Wrought iron was produced by puddling, or boiling, pig iron, often with large quantities of scrap metal in a reverberatory furnace. This process removed some carbon, silicon, and other impurities, resulting in almost pure iron in the form of a pasty mass which was compressed to eliminate slag and then passed through groved rolls under pressure to form muck bars or slabs. In the manufacture of rails, the muck iron was cut into strips, piled in bundles, raised to a welding heat, and again rolled to make a bloom large enough for treatment in a rail mill. The final product was the T-rail, a shape widely adopted by the decade of the fifties, superseding the H- and inverted U-designs which had been used for a time in previous years.[7]

Although superior to cast iron, wrought-iron rails suffered from one important defect. They had a short life because of their inability to withstand sudden shocks, which caused the laminated strips that formed the section to separate. When this happened, the iron could be reprocessed. Scrap rails were heated, rolled into slabs, refashioned as blooms, and finally reformed as rails. While recycling allowed railroads to recapture most of their original investment, it had the effect of raising operating costs.[8]

The shortcomings of wrought iron influenced railroads to look for a substitute metal. The logical choice, in terms of physical characteristics, was steel, but current production methods and high cost ruled against its use. Only small quantities of steel manufactured by the crucible and blister methods were consumed in the fabrication of edged tools, cutlery, and other items where durability justified the expense.[9]

The problem of how to make steel cheaply was resolved by Henry Bessemer. His simple and economical method of transforming molten iron into steel was publicized in 1856, the year that he applied for patents in the United States. Immediately, William Kelly, an ironmaster of Eddyville, Kentucky, claimed priority of invention. Eventually the American won the rights to the pneumatic features of the technique.[10] The resulting controversy between rival claimants discouraged for at least a decade the introduction of the Bessemer process in the United States.

Initial attempts to apply the Bessemer process in England revealed serious technical problems. It was simple, fast, and inexpensive compared to older methods of making steel. In an open-top vessel, a blast of air was forced through molten pig iron, causing slag to form from the oxidation of silicon, other impurities, and a small amount of iron. Carbon also united with oxygen and was eliminated as a gas.[11] After about a dozen minutes, the iron was reduced to an almost pure state. However, when cast in molds, the material did not solidify.

The Bessemer process was made commercially feasible by another Englishman, Robert F. Mushet, who invented a method of removing excess oxygen from the decarburized iron. He discovered that the molten iron, after having been treated pneumatically, could be transformed into a uniform, workable steel by the addition of ferromanganese, better known as spiegeleisen.[12]

Another problem was more difficult to resolve. By accident, Bessemer's experiments with the pneumatic technique had been conducted with molten pig iron that was relatively free of phosphorous. However, when the process was applied commercially in England, that impurity was present in the pig iron, and it was not eliminated by oxidation in the converter. As a result, the steel was brittle and useless.[13]

Once the influence of phosphorous was recognized, the solution was to utilize pig iron that contained very little or none of the objectionable impurity. Since most of the local ores contained large quantities of that element, English iron-makers had to find new sources of the raw material; some of them sought it on the continent of Europe. Eventually, this problem was overcome by the development of the basic lining in the Bessemer converter as a means of removing the pernicious material.[14] Later the basic open-hearth furnace superseded the pneumatic method, because the former was ideally suited for the treatment of molten iron containing phosphorous.

The introduction of the Bessemer process in the United States may also have been delayed to some extent by the fact that known supplies of iron ores, especially in Pennsylvania, were rich in phosphorous. The new method was not applied on a large scale until fields of low-phosphorous ores were opened in the upper peninsula of Michigan. The completion of a ship canal at Sault Ste. Marie in 1855, linking the upper and lower Great Lakes, provided economical transportation for moving bulk cargoes of ores from their sources to industrial communities. After the Civil War, exhaustive tests of the Lake Superior ores revealed their full potential and set the stage for a new era of American economic growth.[15]

The new method of making steel was introduced in the United States by the Kelly Pneumatic Process Company, owner of the American patent. In September 1864, an experimental plant at Wyandotte, Michigan, began making regular "blows" in small converters. Some of that steel was fabricated as T-rails on May 24, 1865, at the North Chicago Rolling Mill.[16] The next day, that demonstration was repeated for a delegation from the American Iron and Steel Association, which was meeting in the Windy City. One witness to that historic event was Robert H. Lamborn, secretary of the trade association and an executive of the Pennsylvania Steel Company.[17] More than a decade later, he was instrumental in establishing the steel rail industry in Colorado.

The rival claims were consolidated in 1866. Because the machinery at the Wyandotte plant infringed on British patents, the Kelly Pneumatic Process Company joined its interests with the American owners of Bessemer's rights, forming the Pneumatic Steel Association for the purpose of licensing manufacturers in the United States.[18] Within a decade, at least 13 firms had been organized, more than one-half of them in Illinois and Pennsylvania. In the former place, the pioneers were the North Chicago Rolling Mill, the Union Steel Company, and the Joliet Steel Company. Early steelmakers in the Quaker State were the Pennsylvania Steel Company, the Cambria Iron Company, the Bethlehem Iron Company, the Lackawanna Iron and Coal Company, the Freedom Iron Works, and Carnegie Brothers & Company. New York, Michigan, and Missouri contained single plants.

Nearly all of the new companies hoped to exploit a potentially large market for Bessemer rails. The Cambria enterprise at Johnstown rolled in August 1867, the first American-made commercial sections from ingots poured at the Pennsylvania Steel Company's converter at Harrisburg.[19] The following month, the Spuyten Duyvil Rolling Mill in New York fashioned T-rails from steel blown in Troy. The total output of steel rails that year was only 2,277 tons.

American railroads were for a time cautious about adopting the new metal sections. The output of iron rails continued to rise, exceeding 800,000 tons in 1872, ten times the steel variety. However, by 1878, the latter had eclipsed the former. Three years later, the volume of Bessemer rails passed 1,000,000 tons; in 1887, the peak year of the decade, the figure rose to more than 2,000,000.[20]

Steel rails were deemed superior to wrought iron in every way except price. Although Bessemer sections cost $174 per ton in 1868, a few lines considered them more economical than iron. Ten years later, the price dropped to $41.50; at that time, about one-fourth of railroad mileage was in steel. The downward-trending price reached $31.50 in 1888, then fell to $26 in the next decade. By the close of the nineteenth century, more than 90 percent of the rail system was constructed of steel, and relics of the iron age were to be found mainly on spurs and sidings.[21]

The demand for steel, most of which was consumed in the manufacture of rails, stimulated the construction of additional blast furnaces, converters, and rolling mills. The industry became the pace-setter for economic growth in the United States. American steelmakers outstripped all rivals, surpassing their British counterparts in 1890.[22] That year, the United States recorded approximately one-third of the world's output of pig iron and Bessemer steel, and American plants were acknowledged models of efficiency and modern technology. However, European mills retained leadership in open-hearth steel until after the turn of the century.

The American Bessemer steel industry experienced two distinct periods of growth in the nineteenth century. The first, from the close of the Civil War to 1873, was halted by a financial panic and depression that ended railroad construction for several years. A revival of that activity in the decade of the eighties led to another round of expansion in heavy industry. From 1880 to 1890, 28 new enterprises were organized, more than half of them to manufacture rails.[23] It was at this time that the first iron and steel plant opened in the West at Pueblo, Colorado.

Precious metals had provided the major impetus for Colorado's settlement and industrial development. The gold rush of 1859, triggered by reports of discoveries near the future site of Denver, had been sustained by wealth extracted from mines throughout the central Rocky Mountains. In the wake of prospectors who roamed, in subsequent years, that rugged terrain, the land previously occupied by the Arapahoe, Cheyenne, and Ute, was claimed for the westward advance of American civilization.[24] Within two decades, the Indians had been restricted to reservations, the San Juan Country to the southwest beyond the continental divide had been opened to fortune-seekers, and the Centennial State, admitted to the Union in 1876, had experienced a major silver boom at Leadville.[25]

A resurgence of mining in the seventies attracted to Colorado people from all parts of the United States and abroad. In that decade, population growth was more rapid than in any previous period except the initial gold rush. Colorado Territory recorded 39,864 inhabitants in 1870; ten years later, the state contained 194,327 residents.[26] In the mineral belt, Leadville experienced the largest increase. Elsewhere, Denver and Pueblo emerged as commercial-industrial centers, serving mining and agricultural regions by means of rail transportation.

A number of railroads were constructed in Colorado during the decade of the seventies. The Kansas Pacific entered Denver from the east in 1870, and in the same year the Denver Pacific linked the territorial capital with the Union Pacific's transcontinental line at Cheyenne. The Denver and Rio Grande built a narrow gauge system from Denver by way of Colorado Springs, Pueblo, and Walsenburg to a point near Trinidad on the Colorado-New Mexico border. The same enterprise opened a line from Pueblo to Canon City and Leadville and commenced work on an extension over La Veta Pass into the San Luis Valley, the gateway to southwestern Colorado and northern New Mexico. The Atchison, Topeka and Santa Fe crossed the plains of southeastern Colorado to Raton Pass, while completing a branch to Pueblo. The Denver and South Park Railroad, with Leadville and Gunnison as objectives, laid narrow gauge tracks through South Park to the Arkansas River. The Colorado Central built westward from Golden to Georgetown and Black Hawk, and northward with Cheyenne as its goal.[27]

In addition to overcoming in a large measure Colorado's geographical isolation from the rest of the nation, railroads encouraged internal development. The new mode of transportation was a major factor in opening the Rocky Mountain region's natural resources and in stimulating a wide range of industrial activites. Denver and Pueblo, for example, became centers for smelting and refining precious metals.[28] As railroad hubs, they had ready access to the coals of Canon City and Trinidad, plus the metallic ores of the mountain camps for treatment in reverberatories and blast furnaces to produce base metals that were by other processes separated into gold, silver, copper, and lead.

A small but important foundry industry also came into being to manufacture iron and other metal products for mines, railroads, households, farms, and ranches. Denver and Pueblo, viewing metallurgical production as the key to economic growth, offered valuable inducements to eastern furnace and rolling mill operators to erect in the West facilities for primary and secondary manufactures.[29]

General William Jackson Palmer and the men associated with him in the Denver and Rio Grande Railway Company planned as early as 1873 to produce foundry iron in the Arkansas Valley. This scheme was blocked by the panic of that year and ensuing depression. However, they acquired through ancillary enterprises possession of large tracts of coking coal, limestone, and iron ore, the requisite materials for making iron and steel.[30] Mineral fuel deposits in Fremont, Huerfano, and Las Animas counties were opened to supply commercial and domestic consumers in Colorado and adjoining states; near Trinidad, ovens were erected to manufacture coke for regional smelters of precious metals.

Conditions in Colorado seemingly justified by 1880 the establishment of heavy industry. That year General Palmer and his associates organized the Colorado Coal and Iron Company for the purpose of building and operating blast furnaces, converters, and rolling mills near Pueblo. At the time, the consumption of foundry iron in Colorado more than equalled the output of a large furnace, while the demand for rails in construction and for renewals more than exceeded the capacity of one of the nation's most advanced rolling mills.[31] Under the circumstances, there appeared to be no reason why iron and steel could not be made as profitably in Colorado as in Pennsylvania — and Pueblo, in the eyes of the promoters, seemed destined to become the "Pittsburgh of the West."

The Colorado Coal and Iron Company, incorporated in early 1880, was a consolidation of the Central Colorado Improvement, the Southern Colorado Coal and Town, and the Colorado Coal and Steel Works companies, all subordinate ventures of the Denver and Rio Grande Railway. The new enterprise brought under the control of a single management the coal lands, iron-ore deposits, limestone quarries, and water resources essential for the production of iron and steel, and the fabrication of commodities from those metals. In addition, the firm obtained from the railroad guarantees of preferential rates for the transportation of raw materials and finished products.[32]

The Pueblo company was for several decades the principal heavy industry in the West. It was the Rocky Mountain region's only producer of iron and steel, as well as the major manufacturer of steel rails, spikes, and track accessories for railroads. In addition, the corporation fabricated a large volume of mine rails, nails, iron pipe, castings, and merchant iron for distribution throughout a large regional market.[33]

The enterprise, as its name suggested, was more than a metallurgical manufacturer; it also owned and operated mines and ovens that supplied coal and coke to commercial and domestic consumers throughout the central Rockies and the western portion of the High Plains, especially in Kansas and Nebraska. When the Pueblo firm merged in 1892 with its chief rival to form the Colorado Fuel and Iron Company, the new concern clearly dominated the regional fuel trade.

The Pueblo corporation — whether as the Colorado Coal and Iron or as the Colorado Fuel and Iron Company — created an industrial base for economic growth in Colorado. The heavy industry attracted capital and labor to the West; founded communities in conjunction with mills, mines, and quarries; encouraged the development of agriculture, and mining of precious metals, secondary manufactures, and transportation; and sustained as a supplier of fuel the vital smelting industry.

Although a promoter of growth, the Pueblo firm engaged in a never-ending struggle to survive as a heavy industry in the West. The steel works, remote from major centers of consumption, served a market that only infrequently required products in volume equal to the enterprise's capacity. The vicissitudes of the metallurgical manufacturer were attributed to competition from technologically more advanced eastern producers, railroad rate wars, and "the waxing and waning" of regional demand that reflected the varying prosperity of mining and smelting of precious metals in the Rocky Mountain region.[34]

For many years, the Pueblo company derived its profits, if any, from the sale of coal and coke; the metallurgical departments, when in operation, usually incurred losses. A major remodeling and expansion program at the turn of the century altered the character of the enterprise at a time when the fuel trade had nearly reached the limits of growth. In the twentieth century, the manufacture of iron and steel products gradually eclipsed the output of coal and coke.

The western company's decision to stress metallurgical production was in part an attempt to redress balance within its own operational structure and also a reaction to changes of far-reaching consequence occurring within the steel industry. The emergence of giant combinations, as evidenced by the appearance of Federal Steel and American Steel and Wire companies in 1898, was carried to conclusion three years later with the formation of the United States Steel Corporation.[35] The first billion-dollar enterprise, the so-called "Steel Trust" significantly altered conditions of production and distribution throughout the country and undermined the Pueblo firm's objective of gaining ascendancy in its own regional market.

At about the same time, the Pueblo company experienced a change in management. The expansion program, completed at least in part by 1903, severely taxed available resources and forced the enterprise to solicit substantial financial support from eastern capitalists. John D. Rockefeller, Jr., and George Jay Gould advanced large sums of money, and, in return for enabling the firm to escape insolvency, they became the most influential members of the board of directors.

Internal reorganization, changing conditions within the metallurgical industry, and new management brought to a close in 1903 an era in the Pueblo enterprise's history.

After that date, the Colorado Fuel and Iron Company remained under the control of non-Colorado interests who struggled to make for a regional manufacturer a secure place in a national market.

The Colorado Fuel and Iron Co.'s Minnequa Works, 1902.

The CF&I Steel Corporation's Pueblo Plant, July, 1975. This photograph is taken from the same spot where the Minnequa Works was viewed in 1902.

2

The Central Colorado Improvement Company

A prime contributor to Colorado's economic growth in the decade of the seventies was the Denver and Rio Grande Railway. That narrow gauge line extended from Denver southward, via Colorado Springs, to the Arkansas Valley. From Pueblo, the company built westward to Canon City, and eventually to Leadville, Gunnison, and Grand Junction. Another branch was constructed from Pueblo to the Purgatoire Valley at El Moro, located a few miles north of Trinidad. From Walsenburg, a midpoint on the southern extension, the road was pushed westward over La Veta Pass into the San Luis Valley, where one division paralleled the Rio Grande River in reaching toward the potentially rich commerce of Santa Fe, and another turned westward toward Durango, by way of Antonito and Chama, New Mexico, to tap the wealth generated by newly opened mines in the San Juan Country.

The promoters of the narrow gauge were for the most part men who had been in the employ of the Kansas Pacific Railroad. When that line penetrated the empty plains of Kansas following the close of the Civil War, it was widely believed that the iron rails would be laid all the way to the Pacific Ocean. With that objective in mind, the company conducted in 1867-1868 surveys of alternate routes to the distant coast along the thirty-fifth and the thirty-second parallels. By 1870, however, ostensibly because Congress withheld assistance in the form of land grants for an extension of the roadbed, the Kansas Pacific terminated in the Colorado territorial capital, connecting with the Denver Pacific Railway, completed earlier the same year to the Union Pacific's transcontinental route at Cheyenne.[1]

Disappointed by the failure of the Kansas Pacific to build westward from the Colorado territorial capital, General William Jackson Palmer, who had supervised construction of the road in western Kansas and eastern Colorado, decided to organize a new company to construct a railroad from Denver southward to the border with Mexico. A man of great energy and broad vision, Palmer was destined to become one of Colorado's leading promoters of railroads and the principal developer of heavy industry at a time when the region was only a short step removed from a wilderness environment.

A native of Delaware, Palmer had grown to manhood in the bustling city of Philadelphia amidst the changing scene of industralizing America. In 1853, at the age of seventeen years, he went to work as a rodman for a surveyor's gang engaged in the construction of the Hemphill Railroad. Two years later, a self-taught engineer, he went abroad to study rail transportation, manufacturing, and mining practices in England and France. While in Europe, Palmer investigated, at the request of J. Edgar Thomson, president of the Pennyslvania Railroad, British methods of burning coal in place of wood as locomotive fuel. On returning to the United States, Palmer served briefly as an officer of the Westmoreland Coal Company, one of the largest mining enterprises of that day, before accepting a position as Thomson's assistant.[2] In the latter capacity, Palmer gained first-hand knowledge of railroad construction and management.

A promising business career was interrupted by the outbreak of the Civil War. As a newly commissioned officer, Captain Palmer raised and commanded a troop of cavalry that served with distinction under General Don Carlos Buell in the Nashville and Shiloh campaigns. In 1862, at the request of William S. Rosecrans, Palmer recruited and trained a regiment which helped turn back Lee's army at Antietam that September, when the Confederates were advancing on Union positions guarding the nation's capital. In the aftermath of that engagement, Palmer was captured and incarcerated for four months before gaining his release through an exchange of prisoners. He returned to the Fifteenth Pennsylvania Cavalry, leading it in the battles of Chickamauga, Missionary Ridge, and throughout the Atlanta campaign. Palmer was brevetted a brigadier general before returning to civilian life at the close of hostilities.[3]

Through the influence of J. Edgar Thomson, who wanted a trusted colleague in a position to look after the interests of eastern capitalists in the Kansas Pacific Railroad Company, Palmer was elected in 1865 secretary and treasurer of the projected transcontinental line. The enterprise, with the aid of federal land grants, was laying track across east-central Kansas. Two years later Palmer commanded a corps of engineers on a survey of tentative routes to the Pacific Ocean along the thirty-fifth and thirty-second parallels. On completing that assignment, he was named manager of construction, in which role he supervised in 1869-1870 the completion of the road to Denver.[4]

Long before the Kansas Pacific reached its destination on the eastern slope of the Rocky Mountains, Palmer and several friends had exhibited an interest in southern Colorado and New Mexico. Following the transcontinental route survey of 1867-1868, he had recommended that the railroad build from western Kansas to Pueblo, over Poncha Pass into the San Luis Valley, down the Rio Grande River valley to a point at or near Albuquerque, and west along the thirty-fifth parallel to California.[5] Palmer also proposed construction of a branch from Albuquerque to El Paso and into Mexico.

Palmer's north-south alternative to the Kansas Pacific's defunct transcontinental line was incorporated on October 27, 1870, as the Denver and Rio Grande Railway Company. That enterprise, according to its Colorado territorial charter, proposed to build from Denver to the confluence of Monument and Fountain Creeks at the base of Pike's Peak, and on to Pueblo in the valley of the Arkansas River.[6] At the latter place, the road was to extend westward to Canon City and South Arkansas, climb Poncha Pass to gain access to the San Luis Valley, and strike southward along the Rio Grande River to the Mexican border. Branches were to link the main line with areas of potential commercial value in Colorado, New Mexico, and Utah.[7]

The leading participants in the Denver and Rio Grande venture were men who, like Palmer, had been associated with the Kansas Pacific Railroad. Charles B. and Robert H. Lamborn had known Palmer since childhood; the former was land officer of the Kansas Pacific, and the latter, formerly an officer in the Fifteenth Pennsylvania Cavalry, was a director of the same corporation as well as an executive with the Pennsylvania Steel Company.[8] Other colleagues from the earlier railroad enterprise included Colonels William H. Greenwood and L. H. Eicholtz, engineers; Captain William F. Colton, surveyor; and Robert F. Weitbrec, accountant.[9] In addition, there was Dr. William Abraham Bell, who had served first as photographer and later as physician for the survey teams that mapped, under Palmer's direction, tentative routes for the Kansas line to the Pacific Ocean.[10]

A Londoner with prominent social connections, Dr. Bell became the salesman for Denver and Rio Grande Railway securities in the British Isles. He had published *New Tracks in North America: A Journal of Travel and Adventure Whilst Engaged in a Survey of a Southern Railroad to the Pacific During 1867-1868,*[11] a book that identified Bell, in the eyes of his countrymen who had money to invest abroad, with the bustling American West. Although Bell resided in Colorado most of his adult life, he frequently returned to his homeland to find investors for the numerous business enterprises in which he engaged with General Palmer.[12]

Other backers of the Denver-El Paso railroad scheme were men of influence in business and politics. Prominent eastern investors included Joseph C. Reiff, formerly the Kansas Pacific's financial agent in Philadelphia; J. Edgar Thomson and Thomas A. Scott, president and vice president, respectively, of the Pennsylvania Railroad; Francis Jackson, Palmer's uncle and an officer of the Westmoreland Coal Company; and William Sharpless Jackson, Palmer's lifelong friend and former treasurer of the Lake Superior and Mississippi River Railroad.[13] Western supporters included Alexander Cameron Hunt, former Governor of Colorado Territory; Wilson Waddingham, land speculator and co-owner of the Maxwell Land Grant; and Jerome B. Chaffee, president of the First National Bank of Denver and also Colorado Territory's delegate to the United States Congress.[14]

The promoters of the Denver and Rio Grande Railway, lacking federal land grants or other large public subsidies, had to rely on private capital to build the first division of the narrow gauge line from Denver to the vicinity of Pike's Peak. Several inducements were offered subscribers to a construction pool of $1,000,000. Contributors received equal amounts of first mortgage bonds with all coupons attached; similar securities on which interest payments were to be deferred two years; the railway's capital stock; a proportionate share of all bonds donated by counties through which the road might pass; and an opportunity to participate in the Mountain Base Investment Fund, an independent land and townsite enterprise.[15] An investor who put $10,000 into the pool was given $20,000 in mortgage bonds, half of which were non-interest-bearing for two years; 100 shares ($10,000) of paid-up capital stock; a claim on a pro-rata share of county bonds; and the right to subscribe to as much as $2,250 in the Mountain Base Investment Fund.[16]

Most of the construction money was raised in Europe. General Palmer and his associates solicited subscriptions in New York and Philadelphia while Dr. Bell canvassed investors in the British Isles. The future of the project was not assured, however, until the Dutch banking house of Wertheim and Gompertz, owner of the Sangre de Cristo Grant in the San Luis Valley of Colorado, agreed to purchase the railway's bonds worth $960,000 at approximately two-thirds of face value.[17]

While the construction company commenced work on the roadbed south from Denver, the promoters turned their attention to the commercial development of land in the vicinity of Pike's Peak. The Mountain Base Investment Fund, controlled by Palmer and his business associates, acquired several thousand acres of agricultural land, townsites, and coal deposits at prices, according to Bell, "which will never again prevail in Colorado."[18] Most of this property was secured from the public domain at the regular price of $1.25 per acre; select parcels or tracts valued for their strategic location or mineral resources were, if necessary, purchased from private holders at much higher costs. The organizers of and investors in the Denver and Rio Grande Railway hoped in this way to profit directly from the appreciation of real estate values that would inevitably result from completion of the first division of the narrow gauge road.[19]

On a tract of ten thousand acres purchased from the National Land and Investment Company, successor to the Mountain Base Investment Fund, Palmer, Robert A. Cameron, former manager of the Union Colony at Greeley, and others organized in 1871 the Colorado Springs Company, which laid out the Fountain Colony, designed to attract settlers to the fertile valley at the foot of Pike's Peak.[20] Town lots were priced from $50 to $200 each, while agricultural tracts sold for from $15 to as much as $25 per acre.[21] Persons of "good moral character" and "strict temperance habits" were invited to reside in the new community. In an effort to attract people of means, for the cost of land precluded the poor, the colonizers advertised throughout the eastern United States and in Great Britain. The response from the latter place was so great that before long, Denver residents called the settlement at Colorado Springs "Little London."[22]

The first division of the narrow gauge was completed on October 27, 1871; on the same day, construction began on an extension to the Arkansas River. The cost of the new construction was underwritten by Wertheim and Gompertz.[23] In addition, residents of Pueblo County approved in a special election the issuance of bonds in the amount of $100,000 to the railroad on condition that its passenger and freight depots were to be located in the town of Pueblo within one mile of the county court house.[24]

The Colorado Springs-Pueblo extension, completed June 19, 1872, initially terminated in Pueblo. However, the depots, yards, and shops were later moved to South Pueblo, a new community controlled by the railroad promoters on the opposite side of the Arkansas River. The Pueblo County Commissioners, concluding that the Denver and Rio Grande had failed to conform to stated conditions, withheld not only the $100,000 in bonds, but an additional $50,000 which had been approved by the electorate to assist construction of a branch from Pueblo to Canon City. This action was later sustained by the territorial courts.[25]

Because of the railroad's actions, a speculative boom in Pueblo triggered by excitement over the approach of the narrow gauge line was shortlived. Faith in the town's future as a commercial-industrial center was momentarily undermined by the transfer of the Denver and Rio Grande's terminal facilities to the south side of the river. Convinced that prosperity was linked to the iron rails, numerous merchants shifted their business establishments to the new community.[26]

Residents of Pueblo looked for a time upon the settlement at South Pueblo as a threat to the very survival of the older town and viewed the railroad's arbitrary action as proof that General Palmer and his associates sought to destroy the established community in order to advance the well-being of its rival. The anguish of the time was expressed by one A. Bach, a local bard who, unlike many of his contemporaries on the north side of the river, realized that the narrow gauge, whether it terminated on one side of the Arkansas or the other, was an instrument of progress that would ultimately benefit both communities and the region.[27]

Old Year
We thought that you were acting queer,
To lead us to believe, you see,
The depot — narrow gauge — would be
Way up the Fountain. Lots we bought,
Because you know, we rightly thought
Lots would go up. They went, for lo,
Who let me ask, could surely know
The depot would be here to-day,
And then to-morrow sneak away,
All right: You've ad[d]ed to the town
Great South Pueblo. Its renown,
E'en now, extended far and wide,
And 'tis our city's boast and pride,
The Nolan Grant, with town and ditch,
Has helped to make the city rich.

The Denver and Rio Grande opened for economic development the agricultural and mineral resources of the Arkansas Valley. General Palmer and the men around him were determined to participate in the region's growth for their own profit and for the benefit of the railroad. The instrument for doing so was an ancillary land company which acquired a vast acreage along the river from Pueblo to Canon City.

The Central Colorado Improvement Company was organized on November 11, 1871, by the men who controlled the Denver and Rio Grande. In anticipation of the extension of the road to Pueblo, the new enterprise purchased a large amount of land in the Arkansas Valley in the hope of buying cheaply and later selling at a handsome profit. The scheme was designed to make money for investors, and to encourage settlement and utilization of the region's resources for the advantage of the railroad whose future depended on ability to generate a large volume of traffic.

When formally chartered under territorial statutes on January 11, 1872, the Central Colorado Improvement Company was authorized to purchase, develop, and sell agricultural lands, mineral springs, coal and iron deposits, quarries, and water rights. In pursuing these objectives, the corporation could establish towns, colonies, manufacturing works, mines, wagon roads, canals, and ditches to irrigate lands in the valley.[28] The incorporators served as the company's trustees. They were General Palmer, William Proctor Mellen, Palmer's father-in-law and a prominent New York attorney; Greenwood, Reiff, and R. H. Lamborn. Mellen was elected president, Lamborn vice president, and S. F. Parrish secretary and treasurer.

The company's officers and others organized the Arkansas Pool in order to raise $1,500,000 with which to purchase land and to conduct a variety of activities. Subscriptions were solicited in the United States, particularly in New York and Philadelphia, and in England. Investors were asked to buy the Central Colorado Improvement Company's bonds, with a generous amount of capital stock serving as an inducement. The purchaser of a $1,000 bond received as a bonus paid-up, nonassessable shares worth $1,750 at par.[29]

Although a total of $1,305,000 was subscribed, only $884,232.29 was paid into the company's treasury. The bonds sold for about 68 percent of par value, enabling the promoters to realize only about two-thirds of the cash they had hoped to gain through the pool.[30]

A major portion of the proceeds were used to underwrite the construction of the narrow gauge railroad to the Central Colorado Improvement Company's coal lands west of Pueblo. The Denver and Rio Grande, unable to obtain funds elsewhere, turned to the ancillary enterprise, which agreed to purchase the carrier's 30-year, 7 percent bonds in the amount of $1,040,000 par value for $825,000 in cash. The railway agreed to build from a point where the Huerfano River emptied into the Arkansas, east of Pueblo, to Canon City. The initial segment, from Pueblo to Labran, was to be completed by May 1, 1872, and the remainder within another two years.[31] For a period of 30 years, or until the bonds were redeemed, the Denver and Rio Grande agreed to transport over its branch line all coal offered at the improvement enterprise's mines at a rate 15 percent less than that charged to other shippers. Subsequently, because the Atchison, Topeka and Santa Fe and the Kansas Pacific lines built to Pueblo from the east, the narrow gauge was not required to extend its line down river from the city.[32]

The Arkansas Valley Branch opened for commercial development land purchased by the Central Colorado Improvement Company west of Pueblo. Along both sides of the river, the firm acquired nearly 84,000 acres suitable, if irrigated, for colonization. South of the future site of Florence, 1,800 acres contained valuable coal deposits. At the head of the valley, 320 acres extended into the "Grand Canon of the Arkansas" and encompassed mineral springs, town lots in Canon City, and a "ditch head" for water-power and canals. Along Grape Creek, a mass of magnetic iron ore called Iron Mountain covered more than a thousand acres.[33]

These properties had been acquired at nominal cost. The company paid $247,411.38 in cash for 86,766 acres, obtaining clear titles. The average cost was about $2.85 per acre, a small price for the best agricultural and mineral land in the valley.[34]

The largest single parcel was 48,695 acres comprising the confirmed portion of the Nolan Grant. Located on the south side of the river, commencing at a point four miles above Pueblo and extending eastward ten miles below the city, the tract ran southward an average distance of about six miles. Although there were small amounts of land in the river bottom, for the most part the property was an undulating plateau traversed by the St. Charles River.

Originally the land had been part of a Mexican grant. Gervacio Nolan, a naturalized Mexican citizen of French ancestry and a resident of Taos, had petitioned on November 14, 1843, Governor Manuel Armijo of New Mexico for a grant of one million acres along the south bank of the Arkansas River on the department's northern boundary. The petitioner claimed the land was needed to "establish a farm capable of furnishing maintenance for my family and a fine range for the raising of large stock and sheep."[35] An executive order dated December 1, 1843, authorized the grant and instructed the Justice of the Peace to mark its lawful boundaries.

Later the same month, Cornelio Vigil traveled to the Arkansas Valley to lay out the Nolan Grant. The limits were fixed by four landmarks: the point where the St. Charles River emptied into the Arkansas; five leagues west of that point along the latter river; to the southwest "half way up the brow of the mountain"; and "following north to south the same brow of the mountains to a point opposite the first mark."[36]

In 1845 Nolan and two associates petitioned for a second grant. Six hundred acres north of the Miranda and Beaubian land in what later became New Mexico Territory was authorized and surveyed. Nolan later purchased the interests of his partners.[37]

Following Nolan's death in 1857, his heirs filed with the Surveyor-General of New Mexico claims for titles to both grants. This action was initiated under the terms of an Act of Congress of 1854 permitting the validation of Mexican land grants authorized prior to the Treaty of Guadaloupe-Hildago. The Surveyor-General recommended in 1861 confirmation of the northern tract only and forwarded pertinent documents to Congress for final judgment.

Seven years later, the House Committee on Private Land Claims finally acted, agreeing to review only one of the heirs' petitions on the ground that Mexican law clearly prohibited one person from holding more than a single grant. Furthermore, the size of the northern tract was restricted to 11 square leagues, to conform to the Mexican laws of 1824 and 1828.[38] The claim, for this reason, was reduced from 3.5 million to 48,695.48 acres.

President Ulysses S. Grant signed into law on July 1, 1870, "An Act to Confirm the Title of the Heirs of Gervacio Nolan, Deceased, to Certain Lands in the Territory of Colorado." The statute specified that "when said lands were so confirmed, surveyed and patented, they shall be held in full satisfaction of all further claims or demands against the United States."[39] The final section of the law provided that upon completion of the survey, the heirs or their representatives were to have three months in which to locate the allotted acreage confirmed by Congress. Failure to do so within the time allowed was to be construed to mean that they had abandoned all claims and that their rights under law had terminated.

By that date, the heirs of Gervacio Nolan had sold their claims to the Colorado grant. They had executed on November 5, 1868, a quit-claim deed to Annie E. Blake in consideration of $10,000. Less than a year later, Mrs. Blake and her husband sold two equal one-third interests in the property to Charles Goodnight, pioneer cattleman, and to Peter K. and Jacob C. Dotson, Pueblo ranchers.[40]

In February 1871, Henry Hodding of Salisbury, acting through Edward E. Whitefield as local agent, entered into an agreement with the Blakes, Dotsons, and Goodnight to sell the Nolan tract at a minimum price of $200,000, for a commission of not less than 5 percent of the transaction. The following month, William Blackmore, an English capitalist who was involved in General Palmer's railroad and land speculations, took over the negotiations for the sale of the property, promising Hodding compensation of not less than $3,000 or equivalent value in land.[41]

The Central Colorado Improvement Company completed arrangements by mid-December for the purchase of the Nolan Grant. The last spark of uncertainty, Bell reported to Blackmore, had been removed, and the fulfillment of the pool's objectives seemed assured.[42]

The promoters selected the allotted 48,000 acres with great care, omitting hills and unfertile areas whenever possible, taking only the best agricultural and mineral lands within the limits of the grant. All but 6,487 acres were chosen from within the 11 leagues approved by Congress; the balance, called "floats," was taken from contiguous lands in lieu of tracts occupied by actual settlers.[43]

The Nolan Grant passed into the hands of the Central Colorado Improvement Company on March 30, 1872. On that day, the Blakes, Dotsons, and Goodnight conveyed to William P. Mellen, trustee, the property for a consideration of $130,000. One month later, Mellen transferred the property to the company for $152,000.[44] There is no record of why the corporation paid more than the trustee, nor is there evidence of what happened to the difference between the two amounts.

Even after the General Land Office finally issued a patent for the Nolan tract on March 5, 1875, the company persisted in its claim to the whole of the original land grant. Two years later, Palmer made an agreement with a Philadelphia lawyer named Theodore Adams who was to seek from Congress the additional 3,450,000 acres once claimed by Nolan's heirs. Adams was to pay all expenses incurred but was to share equally with the Central Colorado Improvement enterprise all lands subsequently acquired within the limits of the original grant.[45] That venture, if seriously prosecuted, was not successful.

The Nolan land, in the hands of the Central Colorado Improvement Company, was offered for sale as townsite and agricultural plots. Approximately a thousand acres were platted in November 1872 as the community of South Pueblo. The "bottoms" along the river became the new town's business center. Forty acres, or about one-third of the area, was reserved for the Denver and Rio Grande Railway's depots, roundhouse, and yards. A variety of enterprises were attracted to the south side of the river because of the availability of transportation facilities.

On the slope leading to the mesa 80 feet above the river, John Blair, a Chicago landscape gardener, laid out Corona Park as the "best class" of residential lots. Circular and semicircular thoroughfares were 100 feet wide and newly lined with young cottonwood trees. Another section for family dwellings was planned south of Abriendo Avenue, a broad boulevard two miles in length that separated Corona Park from the plateau.[46]

South Pueblo and contiguous property were traversed by a canal that carried water from the Arkansas River. The system, completed in 1873, comprised the main watercourse plus ditches and laterals that ran parallel to streets on the mesa, providing water for the 10,000 trees recently planted in the area. The canal cost an estimated $90,000 and traversed approximately 10,000 acres of agricultural land before emptying into the St. Charles River.[47]

On the mesa one mile south of the new town, the Central Colorado Improvement Company constructed a reservoir to furnish water for domestic and agricultural consumption. Linked to the St. Charles by a ditch, Lake Minnequa occupied a natural basin of one hundred and sixty acres, capable of holding about 6,000,000 gallons of water. Because the lake was 42 feet above the highest elevation in the community, water could be distributed by means of iron pipes to all parts of South Pueblo.[48]

The Central Colorado Improvement Company announced in early 1874 that land south of the river opposite Pueblo was to be opened for colonization. Any person of good character was invited to purchase a membership in the colony. Certificates were offered for as little as $50 and as much as $200, with intermediate values at $75 and $150.[49] The price varied according to the value of the property the settler hoped to obtain within the colony.

The rolling mesa was divided into units the size of which increased with distance from the town limits. Bordering South Pueblo, one-acre tracts were priced at from $150 to $200. Farther away, five-acre sites sold for $300, and farms of from ten to twenty acres for as little as $150.[50] The colony reserved land for schools, churches, parks, and other public needs.

Members were permitted to select land within the colony's boundaries. The minimum membership entitled the holder to one or more parcels worth no more than $50. The maximum fee guaranteed the purchaser two plots without regard for assessed costs; for example, a purchaser could choose a business lot and a 40-acre agricultural tract with a combined value of $1,050.[51] Participants had the right to select property at any time, but they had to make one improvement in accordance with their means within six months of location.

The membership certificate was in fact a prepayment, not necessarily complete, for land within the colony. Each participant received a clear title at the close of one year if all regulations had been observed and on payment of any balance due, the difference between the initial fee and the value of the lands occupied.

Special inducements were offered prospective colonists. Those who lived in the East were promised reduced passenger fares and freight rates to South Pueblo. In addition, all unplatted land was available free of charge as open range for livestock until such time as the property was leased or sold by the company.

The South Pueblo Colony was a town development enterprise in fact if not in name. It was designed to expedite sales of lots and to stimulate the growth of the community's hinterland. The promoters hoped to attract several hundred settlers to the area in the spring of 1874, and thereafter an even larger number of people.[52] Immigrants from Michigan, Wisconsin, and Minnesota were expected to seek homes in the warm climate of southern Colorado.

The colonization scheme was at best only moderately successful. While the company's gross earnings from real estate in the years from 1874 to 1876 amounted to about $70,000, there is no record of the proportion that can be credited to the colony venture. Sales of land rose to almost $49,000 in 1874, a time of depression in the United States, and fell to an average of $10,500 the next two years. Income from real estate continued to decline until the end of the decade of the seventies. In eight years, the company derived only $84,380.83 from the sale of all lots and agricultural lands.[53]

Real estate holdings, although concentrated south of Pueblo, extended up the valley into Fremont County. There land acquired from the National Land and Improvement Company and from the public domain contained coal, iron ore, and limestone, all of which the Central Colorado Improvement Company hoped to develop commercially through its own industrial enterprises. The firm's officers also recognized the potential of mineral springs, irrigation, and waterpower resources on the Arkansas River in and near Canon City.

Coal abounded along the eastern slope of the Wet Mountains. Mineral fuel deposits were located south of the river opposite Canon City, and they extended southward about 14 miles while varying in width from about 2 1/2 to 3 miles.[54] Prior to 1872, at least one mine had been opened along Coal Creek where the "Canon Seam" came to the surface or cropped out, affording easy access to an excellent grade of domestic coal. As early as 1867, Thomas T. Slocum had taken up under terms of the Preemption Act a quarter section in that vicinity. Two years later, Joseph T. Musser homesteaded an adjoining 160 acres, on which he opened a small mine. The product was for the most part marketed locally, but some fuel was shipped to Pueblo and to Denver.[55]

The Slocum and Musser tracts were purchased by the National Land and Improvement Company. Dr. Bell, as president of that enterprise, issued a lease on the Musser coal bank to Jeremiah Lester in September 1871. The lessee agreed to pay a royalty of one dollar a ton on all coal taken from the land, and to conduct operations in "the most approved manner of coal mining."[56]

That year, the Denver and Rio Grande announced that it planned to build a branch to the coal field near Canon City. The decision, according to Alexander C. Hunt, reflected the carrier's need for a reliable source of good fuel for its locomotives. Coal produced near Golden was inferior in quality, and the mines at Boulder were overtaxed in supplying the Denver market plus the Kansas Pacific Railroad. The narrow gauge line, if it wanted adequate engine fuel, had to build to the mines in Fremont County.[57]

A branch was constructed to the coal deposits on Coal Creek in 1872 as an extension of the line completed that year from Pueblo to Labran. The Canon Coal Railway, a subsidiary of the Denver and Rio Grande, was incorporated by Palmer, William Wagner, and C. Ford Stevens to build track a distance of three miles to the mine on Coal Creek.[58] Another branch was projected to the head of nearby Oak Creek.

Where the Coal Creek branch left the main line, the Central Colorado Improvement Company bought land for a townsite development under the name of Labran. Because the railway laid track only to that point in 1872, although the right-of-way had been graded to Canon City, the new community seemed destined to become the commercial center of the upper Arkansas Valley. It was also selected tentatively as the location for an iron works which the railroad promoters planned to establish in the proximity of the coal and iron-ore lands of Fremont County.[59]

The new town's dream of greatness was shortlived. Reacting to the improvement company's high prices for lots, James McCandless, who owned extensive tracts of land in the vicinity, organized in 1872 the rival community of Florence. Its growth was rapid. When the Denver and Rio Grande laid track to Canon City in 1874, Labran was doomed to oblivion. By the close of that year, only a railroad warehouse and a few vacant buildings remained, and the hope for a profitable townsite to bolster the incomes of both the Central Colorado Improvement Company and the railway had vanished.[60]

The carrier was accused of having terminated its line at Labran in 1872 in an attempt to foster, to the detriment of Canon City, a commercial center on the Central Colorado Improvement Company's land. While Palmer and his business associates hoped to profit from the townsite, there is no evidence to support the charge that they deliberately sought to destroy the economy of the older community. The narrow gauge had terminated in Labran because there were no funds with which to build beyond that point. Although the improvement company had previously agreed to purchase the railway's bonds in the amount of $825,000, with the proceeds to be used for building the line to Canon City, only $561,084.38 was actually paid to the carrier, leaving a deficit of about a quarter of a million dollars.[61]

The Central Colorado Improvement Company had fallen short of the full payment because its capital resources were less than its own requirements. When the Arkansas Pool had fallen $700,000 short of its goal, the promoters were unable to pay the balance due the railroad. Had they done so, only about $50,000 would have remained for underwriting the development of the Arkansas Valley properties.[62] Since the improvement and railway enterprises were controlled by the same men, the decision to temporarily halt construction of the narrow gauge line at Labran, as they saw it, was in the best interest of both corporations.

The Central Colorado Improvement Company, after purchasing the Musser property in the spring of 1872, decided to open a large mine on the coal outcrop. Machinery was secured to drive a slope at a point readily accessible to the branch railroad from Labran.[63] In July of that year, the firm advertised for contractors to work the property. The lessee was to deliver coal on cars at the tipple. The vein, according to the firm's prospectus, was nearly horizontal with only a slight dip, and the coal was about five feet thick, uniform in quality, free of slate, and easily mined. The owner agreed to prepare the property for an output of 200 tons daily; the operator was to furnish pit cars "and all things connected with the working of the mine . . . and all expenses connected with the mining of coal."[64]

The operation of the mine was taken over in September by John G. Jones, John Burdiss, John R. Rice, Samuel Ingalls, John Jefferies, Joshua Price, J. T. Hackworth, and Alexander M. Cassidy. They agreed to ship coal as soon as tracks were laid to the tipple. Meanwhile, the lessor was to drive an entry, turn rooms, and lay rails to all underground working places. The miners were to receive $2.25 per ton for all lump coal placed on cars, and the daily output was not to fall below 75 tons.[65]

The Canon Mine, as the property was known, shipped in 1873 almost 13,000 tons of coal, or an average of about 43 tons daily. The narrow gauge locomotives consumed about one-third that amount, and the remainder was distributed in the regional market.[66] A second entry was driven later that year to increase production to the rising level of demand. Even then, the mine's capacity was severely restricted by the carrier's inability to furnish an adequate supply of cars, especially during the peak months of fuel consumption from October through April.

The contract with John Jones and his associates was terminated after about one year. Eventually the company paid the partners the sum of almost $4,000 in settlement of all accounts.[67] Thereafter, miners were hired to dig coal on contract at a specified rate per ton, while all other workers in and about the mine were paid daily wages.

The Denver and Rio Grande assumed management of the Canon Mine in 1874, paying the improvement company a royalty of $.25 per ton for all coal taken from the property.[68] This agreement remained in effect until December 14, 1878, the day that the narrow gauge was temporarily leased to the Atchison, Topeka and Santa Fe Railroad.[69]

In turning the mine over to the railroad, the improvement company may have hoped to disassociate itself from a business activity that had aroused widespread criticism, particularly in Pueblo, where the Canon City coal sold for $3.00 more per ton than in Denver.[70] A charge of monopoly was leveled against the company and the carrier during the perennial winter fuel shortages. Even after the improvement enterprise resumed operation of the Coal Creek property, some citizens of Pueblo continued to raise the cry of discrimination. The heart of the problem, according to one critic, was that the coal and transportation companies were one and the same, and the latter refused to carry fuel mined in the Canon district by other firms. This gave the former absolute control over the local market.[71]

In the Denver market, where Canon coal competed with lignite of the northern field, excessive competition and allegedly ruinous prices prompted mine owners and operators to organize in January 1878 the Colorado State Coal Producers' Association for mutual protection and to encourage advances in the art of mining. By agreement among producers, Fremont County coal was to sell for $6.00 per ton and lignite from the Boulder Valley for $4.75.[72] The price differential did not harm the southern product because it possessed superior qualities as domestic and commercial fuel. The combination chose officers from leading firms to enforce the price schedule. David C. Dodge, vice president and general manager of the Denver and Rio Grande, was elected president of the Association even though he had not participated in the organizational meeting.

Before the Coal Producers' Association had a chance to stabilize the region's largest coal market, it was forced to disband. Mass meetings of consumers in Denver condemned the price schedules and threatened to boycott any company that attempted to enforce them. Dodge, speaking for the railway, quickly disavowed the combination. Other operators, fearing the public's wrath, followed suit.

The Association's demise may have been of little concern to the Central Colorado Improvement Company for it was hard-pressed to expand production at Coal Creek fast enough to keep pace with demand. New hoisting machinery was installed in 1878 to increase the output to 600 tons daily. That year, the enterprise owned 2,300 acres along Coal and Oak Creeks and had halted production from leased land.[73]

Coal mining was the Central Colorado Improvement Company's most profitable business activity. In eight years, it invested $47,987 in equipment, including homes for miners and their families, and purchased an additional 2,240 acres of coal land in Fremont County.[74] During that time, production amounted to 270,390 tons. The return equalled $82,000, or about $.30 on each ton of coal mined.[75]

Originally, coal mining was to have sustained another industry. Long before the narrow gauge reached the Arkansas Valley, Palmer and his associates had planned to manufacture iron from raw materials that abounded in Fremont County. In the summer of 1871, Robert Lamborn sent to his friends in Colorado Territory analyses of samples of magnetite taken from Grape Creek southwest of Canon City. Tests conducted by chemists of the Pennsylvania Steel Company revealed a high percentage of metal, more than enough to justify the erection of a blast furnace, rolling mill, and foundry. Whether from ignorance or excessive enthusiasm, the promoters did not initially view substantial amounts of impurities in the iron ore as a deterrent to their projected enterprise.[76]

Several months later, however, Robert Lamborn warned Mellen to avoid a hasty investment in an iron works. The Philadelphian argued that the promoters should first build the railway up the Arkansas River and construct a wagon road to the iron-ore deposit. Then they should erect a rolling mill at the mouth of the "Grand Canon" where waterpower was readily available. That facility should be supplied with iron, not from a blast furnace but from a charcoal forge, for which fuel was easily accessible from local pinon forests.[77]

Canon coal, Lamborn cautioned, had not been tested to determine if it would make a solid coke for the manufacture of iron, or if it could be employed in raw form in the smelting process. Furthermore, Grape Creek ore was untested, and experiments elsewhere indicated that for good results magnetite had to be mixed with hematite, which was not known to exist in the region. Finally, Lamborn suggested that from 80 to 100 tons of the ore should be sent to St. Louis for tests in the furnaces of the Carondelet Works.

Subsequent tests of Canon coal justified

Lamborn's caution. A Philadelphia chemist and metallurgist reported in March 1872 that the fuel did not form a coherent coke.[78] Hope for the projected iron furnace rested, therefore, on the possibility that the coal could be used directly with iron ore in the reduction process.

Even though the quality of the raw materials remained unknown, the Central Colorado Improvement Company, acting through A. C. Hunt, opened negotiations in February 1872 with Colonel James Archer, a Denver businessman, for the erection of an iron works at or near Canon City. As an inducement, Hunt proposed that the entrepreneur be given free access to 40 acres of iron ore, 100 acres of coal, plus a building site on the river above Canon City.[79] While this suggestion was not adopted, the corporation's proposition was generous. It guaranteed Archer the right to take from Grape Creek without charge for a period of five years all ore required "for furnaces, foundries, rolling mills and other works to be erected . . . on the property of the company, at such place or places near Canon City as may be mutually satisfactory and advantageous." For coal mined on the enterprise's land, Archer was to pay the nominal royalty of $.025 per ton.[80]

Archer declined the proposition because the company failed to guarantee rail connections to the sources of iron ore and coal, and to fix reasonable rates for the transportation of those materials. He insisted that the royalty on coal should be for a period longer than five years, that unless the corporation offered not only "liberal inducements for the present but security in the shape of cheap coal & iron railroad transportation," there was nothing to be gained from undertaking the risk of raising capital for the proposed enterprise.[81]

Despite this setback, the company did not abandon hope for an iron works in the Arkansas Valley. R. Neilson Clark, chief engineer for the Coal Creek Mine, suggested in 1874 that Pueblo was the ideal site for a metallurgical industry.[82] That theme was elaborated in the corporation's annual report of 1876, which assured stockholders that a rolling mill would be erected at South Pueblo, a point midway between the coking coal and carbonaceous iron ores of Trinidad and the magnetite of Grape Creek.[83] The promoters had recognized the shortcoming of Canon coal but continued to labor under the impression that Fremont County iron ore was suitable for large-scale blast furnace operations.

Meanwhile, the Central Colorado Improvement Company commercially developed the thermal springs at Canon City. The Hot Springs near the entrance to the "Grand Canon of the Arkansas" had been purchased in 1872,[84] but was neglected until shortly before the Denver and Rio Grande completed its line to the city in the summer of 1874. A health resort was established on the south bank of the river about a mile from the center of the town. The water's mineral content included sodium chloride and carbonate, lime and magnesium carbonate, and sodium phosphate and iron, making the spring an ideal bathing spot for invalids and persons afflicted with "rheumatic and cutaneous diseases."[85]

Perhaps as a method of generating traffic for the railroad, the Central Colorado Improvement Company erected in the summer of 1874 a bathhouse at the springs. The 90-degree temperature of the water, according to one account, could remove "dirt from the hide of the most obstinate bull whacker," leaving his complexion "as rosey and fair as that of an infant."[86] Victims of rheumatism and all kinds of skin ailments were invited to seek relief in the medicinal waters at a cost of $.50 admission per person.

Over a period of four and one-half years, the Hot Springs earned only a small profit. The property was leased to Dr. J. W. Dawson for perhaps two years, after which J. P. Ring managed the spa as the "Canon City Warm Springs and Bath House."[87] Net earnings, as of January 23, 1880, when the site passed to the control of a successor company, amounted to less than $1,000.[88]

This inability to generate profits was reflected in nearly all of the Central Colorado Improvement Company's business activities. Although active from 1872 to early 1880, the corporation failed to pay dividends to stockholders who had been led to expect large returns on their investments. Originally, the corporation had stated that income from the Denver and Rio Grande's bonds on the Arkansas Valley Branch and from the Coal Creek Mine would be more than enough to pay the fixed charges on the $1,500,000 of bonds issued in 1872.[89] Income from additional coal mines, iron mines, irrigation and waterpower resources, townsite and agricultural lands would produce, it was said, large surpluses for distribution to security holders.[90]

That hope was shattered by the Panic of 1873. The economic reversal was a "heavy and long-continued shock to capital upon the development of a perfectly new country."[91] The Denver and Rio Grande temporarily abandoned plans for an extension south of Pueblo; the Kansas Pacific and the Atchison, Topeka and Santa Fe, broad gauge lines building up the Arkansas Valley from the East, suspended construction after defaulting on interest payments to bondholders. The influx of settlers into the region declined and the South Pueblo townsite ceased to grow.

Before the improvement company had a chance to recover from the initial impact of the Panic, news arrived in the West of Mellen's death. He had suffered a cerebral hemorrhage in Cincinnati in mid-November.[92] On the first day of the following month, the board of directors elevated Palmer to the presidency. In the following five years, he struggled to maintain the corporation's solvency and that of the railway while the economy of Colorado faltered under the impact of a general depression.

During that time, Palmer insisted on expanding the Denver and Rio Grande. The Arkansas Valley Branch was completed in 1874 to Canon City; two years later, the southern extension reached El Moro, a new town near Trinidad.[93] This additional construction in the midst of depression severely strained the narrow gauge's resources.

Impending disaster became a reality on May 1, 1877, when the Denver and Rio Grande failed to pay interest to bondholders and declared its inability to meet similar obligations at six-month intervals during the coming year. The improvement company, because of its large holdings of the railway's bonds, had no alternative but to default on its own bonds.[94] In order to avoid insolvency for both enterprises, the Central Colorado firm accepted certificates aggregating $109,200 for three payments and issued its own paper in the amount of $116,010. That policy was momentarily resisted by English security owners who, at a meeting in London, appointed a committee to investigate the financial condition of Palmer's corporations.[95] Eventually, all bondholders accepted five-year certificates bearing six percent interest annually. Both companies resumed service on their debts before the close of 1878, and all certificates were redeemed within three years.[96]

In the summer of 1878, the Denver and Rio Grande modified and extended its contract with the Central Colorado Improvement Company for transportation services. The latter was again guaranteed for 30 years a rate 15 percent less than that charged other shippers of coal on the Arkansas Valley Branch. The railway also agreed to provide for a similar period annual passes to the Pueblo firm's officers, franking privileges on telegraph lines, and all "tracks, spurs, sidings and switches" required for the operation of coal mines in Fremont County. However, none of these concessions were to apply to affiliated enterprises. In return, the carrier gained the right to purchase fuel for its own use at a price not to exceed $2.00 a ton.[97]

This agreement was vital to the improvement company because of its great dependence on coal mining for profits. However, despite persistent financial problems, the corporation never abandoned its hope for an iron works in the Arkansas Valley. In 1878, a tumultuous year when the Denver and Rio Grande was embroiled in a bitter fight with the Santa Fe over an extension from Canon City to Leadville, the narrow gauge line contemplated a branch to the new boom town of Silver Cliff. Charles Lamborn informed stockholders that in penetrating the Wet Mountains the carrier would open for commercial development the rich iron deposit along Grape Creek. Ore from that site would assure the availability of raw materials for the erection and operation of a blast furnace and rolling mill in South Pueblo.[98]

Although the Grape Creek line was not

constructed at that time, Palmer and his associates decided in 1879 that the time was right for the establishment of an iron and steel industry in southern Colorado. In order to achieve that objective, it was necessary to consolidate their nonrailroad ventures as a single corporation. Therefore, on December 13, 1879, stockholders of the improvement company approved a plan for merger with the Southern Colorado Coal and Town Company and the Colorado Coal and Steel Works Company. On January 23, 1880, after shareholders in the later enterprises had endorsed similar schemes, all three firms were united as the Colorado Coal and Iron Company.[99]

The Central Colorado Improvement Company closed its business operations with a deficit. It had derived from real estate, securities, mining, and other ventures in eight years a gross income of almost $700,000, while expenses, including taxes and interest on bonded indebtedness, exceeded $824,000. The enterprise ceased operations on January 22, 1880, with a debt of more than $120,000 and a record of having paid no dividends to stockholders.[100]

Nevertheless, the company had numerous achievements to its credit. It had hastened the settlement and development of the Arkansas Valley. South Pueblo was a thriving commercial center with a population of several hundred inhabitants, and a sizeable proportion of that community's hinterland was irrigated and under cultivation. Fremont County's coal deposits had been opened on a large scale, providing not only railroads but also domestic consumers throughout the region with an excellent fuel. The iron ore of Grape Creek, although untapped, had sustained a dream that was to become a reality within less than two years.

Fremont County Camps

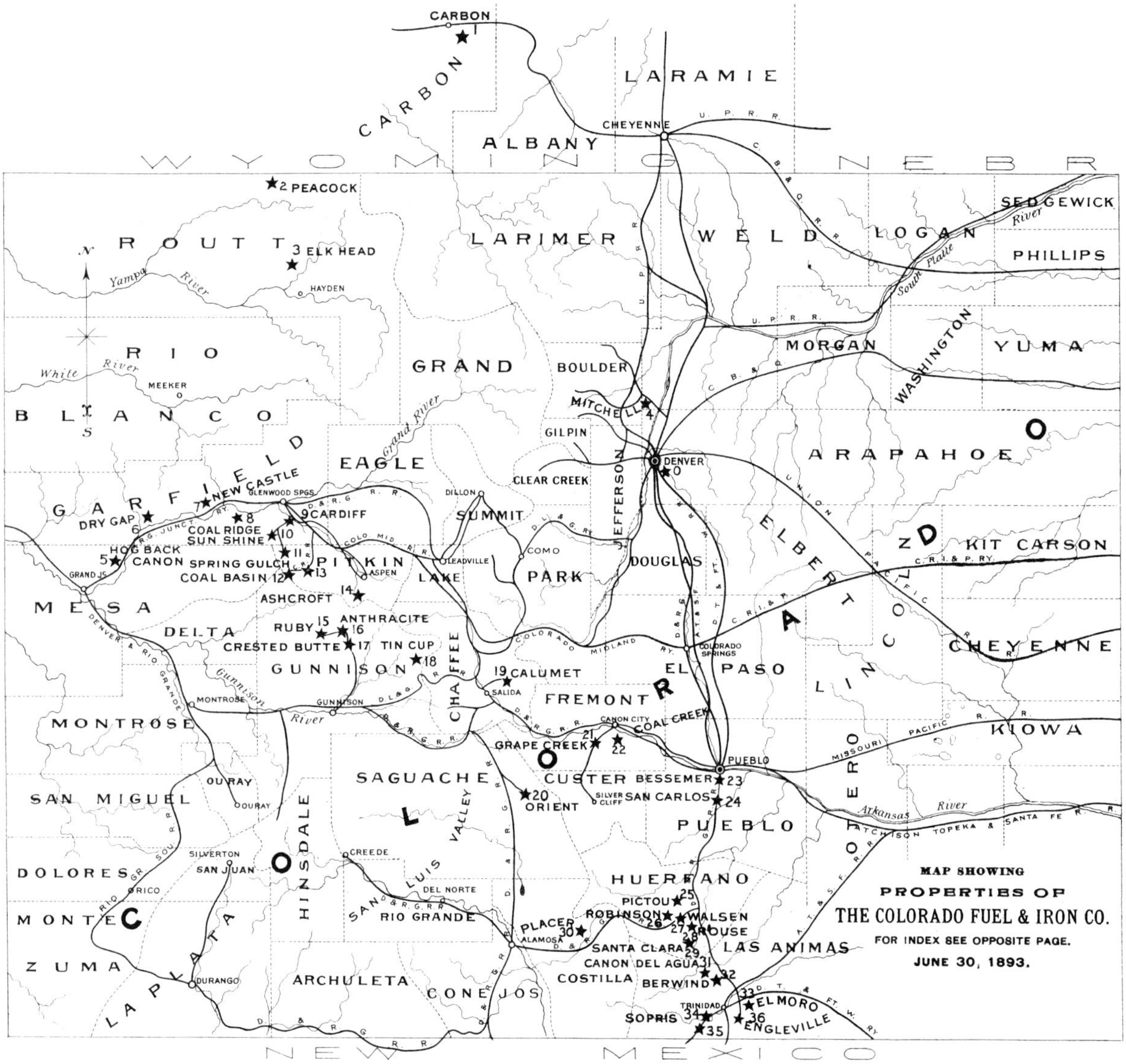

Properties of The Colorado Fuel and Iron Co., June 30, 1893: (0) General Office in Denver, (1) steam coal lands in Carbon County, Wyo., (2 and 3) coal deposits in Routt County, (4) Mitchell Mine in Weld County, (5) coal deposits in Mesa County, (6)Dry Gap coal lands in Garfield County, (7) Newcastle Mine, (8) Coal Ridge Mine, (9) Cardiff coke ovens, (10) Sunshine [Sunlight] Mine, (11) Spring Gulch Mine, (12) Coalbasin Mine, (13) Redstone coke ovens, (14) Ashcroft iron ore deposit, (15) Floresta Mine, (16) Anthracite Mine, (17) Crested Butte Mine and coke ovens, (18) Tin Cup iron deposit, (19) Calumet iron mine, (20) Orient iron mine, (21) Grape Creek iron ore deposit, (22) Coal Creek Mine, (23) Bessemer Works, (24) San Carlos lime quarry, (25) Pictou Mine, (26) Robinson Mine, (27)Walsen Mine, (28) Rouse Mine, (29) Pictou Mine, (30) Placer iron ore mine (closed), (31) Canon del Agua coal deposit, (32) Berwind Mine, (33) El Moro coke ovens, (34) Sopris Mine and coke ovens, (35) Smith Canon coal deposit, (36) Engleville Mine.

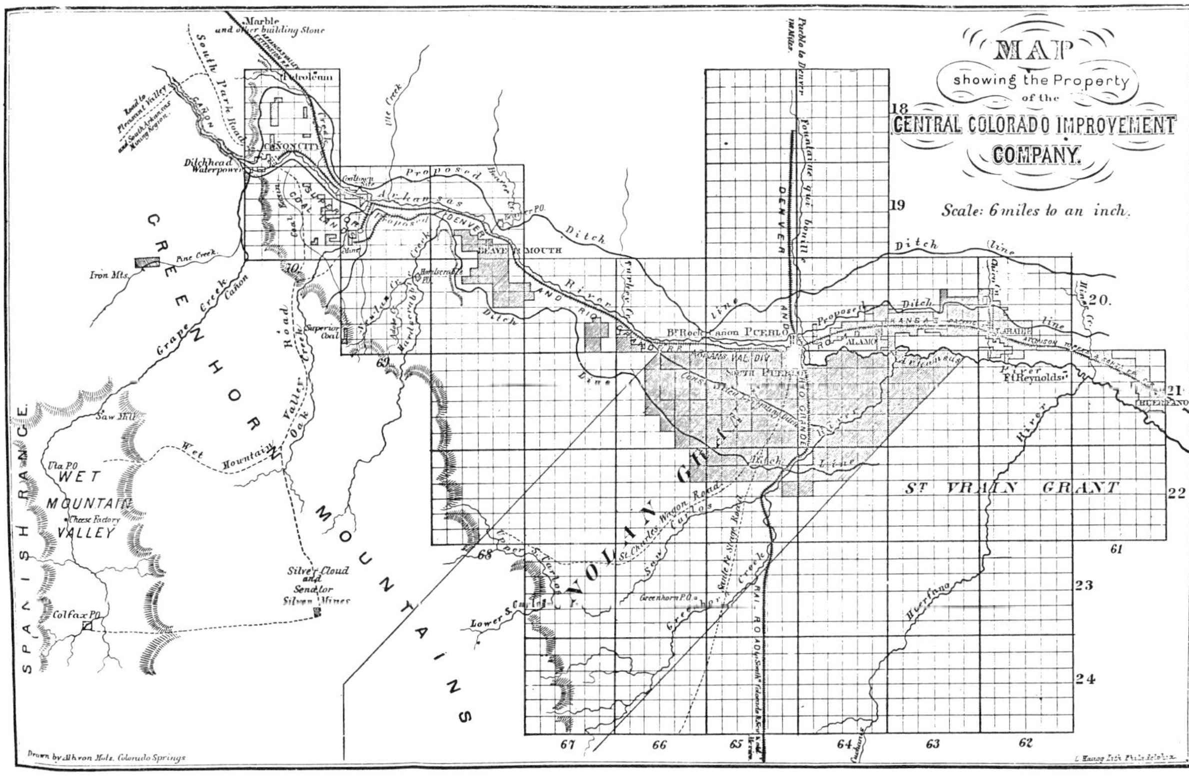

Properties of The Central Colorado Improvement Co., 1876.

General William Jackson Palmer (1836-1909). The founder of the Denver and Rio Grande Railway and the City of Colorado Springs, Palmer was the principal organizer and president of the Colorado Coal and Iron Co. from Jan. 23, 1880 until he resigned that position on April 21, 1884. His successors were Henry E. Sprague (1884-1889), Edward J. Berwind (1889-1892), and Channing F. Meek (1892).

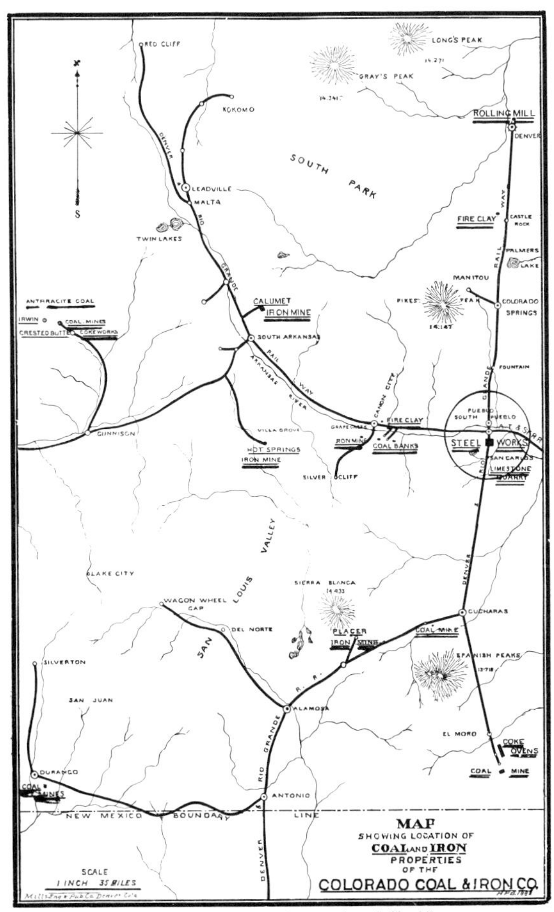

Properties of Colorado Coal and Iron Company underscored with red lines.

Colorado Coal and Iron Co. coal and iron properties, ca. 1885.

Rockvale Mine tipple (Fremont County). The camp was established in 1879 by the Canon City Coal Co., a subsidiary of the Atchison, Topeka and Santa Fe Railway. In 1896, that company leased all of its coal mines in Colorado to the Colorado Fuel and Iron Co. The Pueblo firm ceased operations at Rockvale in 1927, by which time it had extracted 4.5 million tons of coal.

Colorado Supply Co. store at Rockvale.

Coal Creek No. 2 slope, one of three mines operated at Coal Creek, Fremont County, by the Colorado Fuel and Iron Co., and its predecessor, was also known as the Canfield or Caldwell mine. This property was operational from 1884 to 1934, most of that time under the direction of the Pueblo corporation, and yielded 4.3 million tons of bituminous coal.

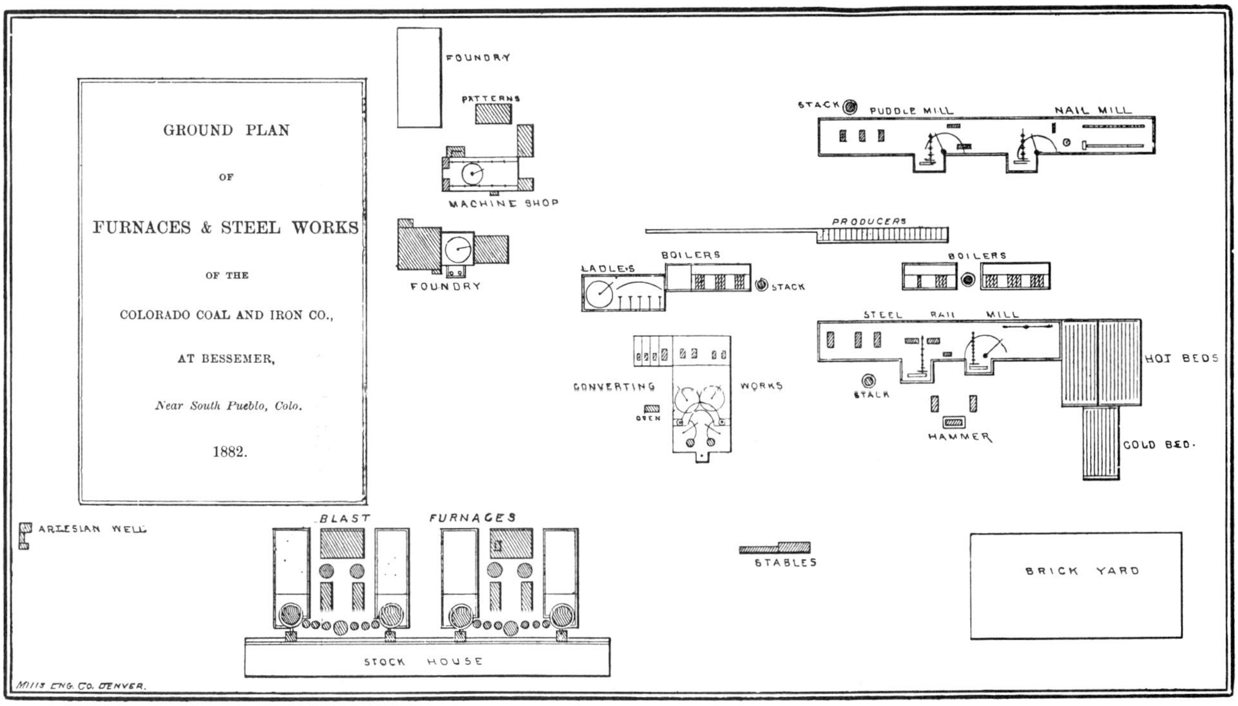

The Colorado Coal and Iron Co.'s iron and steel plant at Bessemer, south of Pueblo, 1882.

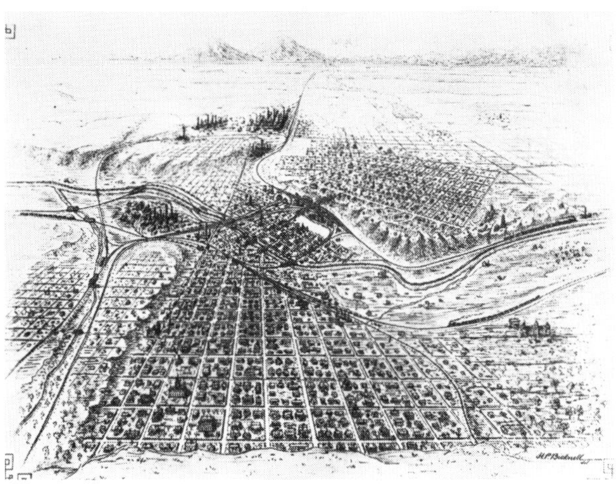

Bird's Eye View of Pueblo, 1888. The Colorado Coal and Iron Co.'s Bessemer Works is visible at the top and center, with South Pueblo to the upper-left. The circular drives of Corona Park are recognizable on the bluff south of the Arkansas River. Courtesy of The State Historical Society of Colorado.

Fremont Mine, better known locally as the Bear Gulch shaft, was purchased by the Colorado Fuel and Iron Co. in 1896. The plant was rebuilt in 1905 after a fire destroyed all surface structures. Production continued until 1927, and the mine yielded for the Pueblo firm 2.6 million tons of coal.

Miners in front of the Colorado Fuel and Iron Co.'s office at the Fremont Mine, 1914.

Rolling mill gang, possibly the crew for the 20-inch mill, ca. 1890. Colorado Coal and Iron Co., Pueblo Works.

Bar-Mill gang, ca. 1890. Colorado Coal and Iron Co., Pueblo Works.

3

The Southern Colorado Coal and Town Company

The men who controlled the Denver and Rio Grande organized the Southern Colorado Coal and Town Company to assist the narrow gauge line in building an extension to the vast coal resources of the Cucharas and Purgatoire valleys. Therefore, the new enterprise performed with respect to the railway a role similar to that played in previous years by the Mountain Base Investment Fund and the Central Colorado Improvement Company.

The Southern Colorado Coal and Town Company was created May 6, 1876, by William S. Jackson, Robert F. Weitbrec, and William Wagner, all of whom held executive positions with the Denver and Rio Grande Railway Company. Ten days later the same men filed articles of incorporation with the territorial government in Denver. The enterprise, by terms of its charter, was authorized to raise a capital of $1,000,000 for the purpose of acquiring, developing, and selling townsite and coal lands in Huerfano and Las Animas counties.[1] General Palmer, the Lamborn brothers, and Jackson comprised the first board of directors. Henry McAllister, Jr., was elected president (a position later occupied by Palmer), C. B. Lamborn, vice president, and W. B. Gaskill, secretary and treasurer. All of the officers were committed to a policy of aiding the Denver and Rio Grande in constructing a southern extension.

In order to build south of Pueblo, Palmer had initially sought assistance from the federal government. On March 4, 1872, Jerome B. Chaffee introduced in the United States House of Representatives a bill to aid the construction of a narrow gauge line from Denver to El Paso by permitting the Denver and Rio Grande to purchase alternate sections from the public domain at a price of $1.25 per acre.[2] While that proposal died in the Committee on Public Lands, Congress granted a right-of-way to the projected north-south line. By terms of a law of July 8, 1872, the government agreed to relinquish 100 feet on either side of the road bed plus 20 acres for each 10 miles of track completed to serve as locations for stations and shops. To qualify for these benefits, the railway had to be constructed to Santa Fe within five years.[3]

In 1873, the narrow gauge line solicited aid from foreign investors to underwrite construction from Pueblo to Trinidad. A financial panic that year, widespread railroad bankruptcies, and the scandals of the Union Pacific's Credit Mobilier defeated the proposal in England.[4] The following year, James C. Parrish tried unsuccessfully to gain support from Dutch as well as English bankers. Palmer also attempted to refinance the Denver and Rio Grande and to fund the southern extension through a new pool, the Franco-American Guarantee Company, composed of French bondholders. However, they were no more eager than capitalists in Great Britain and the Netherlands to invest in a railroad venture during a time of economic dislocation in America.[5]

Unable to raise money in Europe, Palmer revived a technique he had successfully employed on two earlier occasions. He offered to contributors to the southern extension a combination of Denver and Rio Grande securities and a potentially profitable land investment. He was able to do this because two groups of land speculators with large holdings in southern Colorado were willing to cooperate with the railway. The projected Pueblo-Trinidad line would substantially increase the value of their investments.

Acting as agent for the Denver and Rio Grande, Bell in 1875 approached Ernest P. Stephenson and William S. Jackson, trustees for parties who controlled about 13,000 acres in Huerfano and Las Animas counties. They agreed to place their holdings in a new enterprise, the Southern Colorado Coal and

Town Company, and to surrender one-half of the capital stock to the railway.[6] Bell promised to arrange for the extension of the narrow gauge line to a point on or near their properties and to obtain preferential freight rates on all coal and coke shipped by the corporation. On the same day, he formalized these concessions in an agreement with General Palmer as president of the railroad.

Also on that day, the Denver and Rio Grande negotiated with the Union Contract Company for the construction of track from Pueblo to Trinidad. The latter firm guaranteed completion of the line by December 1, 1876, in return for the railway's first mortgage bonds in the amount of $1,925,000 and $2,500,000 of its fully paid capital stock.[7]

A construction pool was offered to investors. In the summer of 1875, General Palmer personally offered the plan in Philadelphia and New York while Dr. Bell returned to England to seek investors in the railway's 120-mile extension.[8]

A "confidential" subscription paper outlined the benefits of the proposed extension south of Pueblo. It pointed out that the coal of the Trinidad field not only made an excellent coke which was needed by smelters of silver and other precious ores, but that southern Colorado possessed abundant iron ore and limestone which, with the available fuel, justified the erection of at least one large furnace and rolling mill.[9] Furthermore, the manufacture of iron there was protected by a $20.00 per ton freight tariff from the nearest mill in St. Louis.

In addition, by means of the southern extension, the Denver and Rio Grande would exploit the so-called "Santa Fe trade." The railway, from a terminus in or near Trinidad, would carry northward to trunklines at Pueblo and Denver large quantities of goods freighted from New Mexico and then transport in return merchandise imported by that region.

There was also a potentially profitable trade at Cucharas. At that point, merchandise and equipment would be forwarded by freighters to the San Juan Country, and receipts of hides, cattle, bullion, and ores would be shipped by rail to Pueblo and Denver. The continuing migration into southwestern Colorado and the growth of mining enterprises in that region promised a large volume of traffic for the southern division of the narrow gauge.

Finally, the extension would open a large traffic in coal from the Cucharas and Trinidad fields. The former was known as a superior metallurgical fuel,[10] while the latter was already consumed in forges, foundries, and gas works as far away as Denver.

Subscribers were offered a combination of securities. For each $750 paid into the pool, an investor received $1,000 of the Denver and Rio Grande Railway's 7 percent gold bonds, $1,000 of the same company's capital stock, and $300 of the Southern Colorado Coal and Town Company's shares. At the current gold premium the bonds alone, according to the promoters, were capable of returning 11 percent per annum.

No subscription was binding until $400,000 had been pledged, enough to build the narrow gauge line to the Cucharas River. One-fourth was payable on receipt of notice from the railway company, and the balance was to be called at intervals from July 1 to December 31, 1875, as required for construction.

The pool was fully subscribed. Contracts for grading south of Pueblo were awarded in November of 1875, and track-laying commenced the following January. The line was completed to Cucharas on February 22, 1876, and to El Moro, a new town four miles from Trinidad, on April 6 of the same year.[11]

In May, the railway promoters formally organized the Southern Colorado Coal and Town Company. That month, Stephenson conveyed to the enterprise approximately 4,000 acres in exchange for capital stock worth $320,700, one-half of which was assigned to the construction pool.[12] Also, Jackson transferred 9,000 acres in return for shares amounting to $678,800, one-half of which were applied to the subscription fund for the Trinidad extension.[13] The corporation subsequently purchased for $73,700.34 additional land controlled by Jackson as trustee.[14]

The land conveyed by Jackson had been acquired as a speculative venture by men who anticipated that construction of the narrow gauge south of Pueblo would cause an appreciation of coal, iron, townsite and other properties in Huerfano and Las Animas counties. In 1873, the speculators formed the Trinidad Pool to obtain land from the public domain.

The key personnel in that organization were Alexander C. Hunt, right-of-way purchasing agent for the Denver and Rio Grande, and Jackson, secretary and treasurer of the railway. Hunt obtained titles to land, more often receiver's certificates for preemptions, for which Jackson, as trustee of the pool, paid him from $500 to $800, depending on location.[15] In all, some 61 parcels, aggregating about 10,000 acres, were acquired in this way.

A total of 84 shares in the pool netted the trustee $46,500. Fifty-one were for $500 each and thirty-three for $600. With that money, Jackson purchased titles or certificates which he later transferred to the Southern Colorado Coal and Town Company, taking in return fully paid capital stock worth $399,400. Unexpended funds and the town company's shares were distributed to pool members in proportion to their contributions.[16]

Each share, worth $3,844 in the coal and town company's stock, was assessed by prior agreement a trustee's commission equal to 10 and 20 percent, respectively, on the two categories of investments in the pool. After that deduction, the new corporation's shares were worth $3,460 and $3,076.[17] The trustee received altogether $45,475 in stock which could not earn dividends until each subscriber had been fully repaid in the cash settlement and subsequent profits an amount equal to the sum invested in the pool.

The trustee's settlement with each shareholder included a refund of $276.78, money that had not been used to purchase land. The investors, therefore, had to earn either $223.22 or $373.22, depending on the shares they held, before Jackson was entitled to participate in the earnings of the Southern Colorado Coal and Town Company.

Members of the Trinidad Pool, according to Jackson, numbered about 50 persons. From extant records, it is possible to identify 32, about one-half of whom were closely associated with either the Denver and Rio Grande Railway or the Central Colorado Improvement Company. The remainder were principally residents of Philadelphia and New York, and many were connected with railroads and the iron and steel industry.[18]

The extension of the Denver and Rio Grande south of Pueblo opened those lands for commercial exploitation. Initially, the coal and town company laid out two communities, one where the railway struck the Cucharas and another near Trinidad. Cucharas was platted east of Walsenburg. In September 1874, that community consisted of four houses and an irrigation ditch had been completed to support agricultural endeavors. Several persons had been attracted to the region "with a view of making a permanent settlement."[20]

However, the town was practically abandoned even before the railway reached that place in February 1876. The coal and town company had by then decided to open coal deposits near Walsenburg, and the Denver and Rio Grande was subsequently extended in that direction.[21]. Eventually, the buildings at Cucharas were moved to the new town of La Veta on the narrow gauge's extension to the San Luis Valley.[22]

El Moro was laid out at the Denver and Rio Grande's terminus in the Purgatoire Valley. Residents of Trinidad viewed the new town, four miles distant, as an attempt by the coal and town company, abetted by the railway, to supplant their community as the commercial center of the region.[23]

While the decision to halt the narrow gauge on the north side of the river may have been prompted by a desire to enhance the value of the coal and town company's real estate, the official reason for doing so was that titles to lands within the old town were in doubt because of the unadjudicated Lieterdorfer Grant.[24] It is also possible that the founders of El Moro believed that Trinidad, because of its topography, was incapable of further growth. James T. Gardner, a geographer with the United States Geographical Survey, had suggested the previous year that the railway, for that reason, should locate its depot and yards on the broad plateau north of the Purgatoire.[25]

El Moro prospered as a distributing center for extreme southern Colorado and northern New Mexico. The town's monopoly on the movement of freight in and out of the region, although challenged by the Kansas Pacific's terminus in La Junta, was not ended until 1879 when the Atchison, Topeka and Santa Fe Railroad reached Trinidad.[26] That year, the new community's commercial activity declined as mercantile houses, following the rails in search of larger markets, transferred operations to the south side of the river. Thereafter, El Moro remained an important shipping point for the narrow gauge line and the site for the coal and town company's coke ovens, but it clearly lived in the shadow of Trinidad.

The townsite venture in Las Animas County was not financially successful. The company incurred expenses amounting to nearly $3,300 in developing the community and derived only $3,400 from the sale of 100 lots. The Denver and Rio Grande, as a silent partner in the enterprise, may have been a heavy loser. According to one estimate, a boycott of the narrow gauge line by Trinidad merchants in the spring of 1876 cost the carrier $15,000 in diverted freight alone. Its total loss may have been much greater.[27]

The Southern Colorado Coal and Town Company inaugurated mining operations at Walsenburg, a settlement adjacent to the ford of the Cucharas River on the trail from Pueblo to Trinidad. Originally known as La Plaza de Los Leones, in honor of Alcalde Don Miguel Antonio Leon, the community was renamed for Fred Walsen, a German immigrant who opened a trading post at that place in 1870.[28] Two years later, it was designated the seat of government for Huerfano County. Its rapid growth commenced with the arrival of the narrow gauge and large-scale exploitation of the area's coal resources.[29]

Alexander Cameron Hunt and a man named Allen opened a mine one mile west of the town in March 1876, before the land was formally conveyed to the Southern Colorado Coal and Town Company. They leased the property for a period of two and one-half years, selling the output to the railway.[30]

Twenty-five miners opened an eight-foot seam. The output, initially only 16 tons daily, was transported by wagon to the railway at Cucharas until a branch was completed to Walsenburg.[31] The product, an excellent steam and domestic fuel, was consumed by the Denver and Rio Grande's locomotives, and that line carried a large volume northward to markets in Pueblo and Denver.[32]

The coal and town company assumed management of the mine in October 1878, at which time its capacity was enlarged. Superintendent Charles Lamb increased the output to 300 tons daily by driving new entries and employing additional miners.[33]

Farther south, the company opened, in the autumn of 1877, a mine in the coking coal field of Las Animas County. On the site of a coal bank previously worked for the wagon trade by W. G. Riffenberg, a local merchant, parallel drifts were driven into an outcrop at the base of Fisher's Peak four miles southeast of Trinidad and six miles south of El Moro at the end of a newly constructed branch of the narrow gauge.

Although officially known as the El Moro Mine, it was popularly identified with George U. Engle, the first superintendent who retained that position for six years. The name was altered officially to Engleville in 1892, making the mine conform to the designation of the adjacent town; three years later, both were called Engle.[34]

The coal seam, eight to ten feet in thickness, was comparatively level, free of gas in appreciable quantities, and adapted to large-scale production. Hunt and Allen, contractors, commenced shipments in December 1877. Four months later, the company took over direction of the mine.[35] All screenings and some slack was sent to the ovens at El Moro for the manufacture of coke. Lump coal was sold to the Denver and Rio Grande and shipped to Pueblo and Denver to fill the demand for metallurgical fuel.

The Southern Colorado Coal and Town Company was the largest producer of coal in the region. Its Walsen and El Moro fuels were the only products of Huerfano and Las Animas counties to enter the Pueblo and Denver markets in appreciable quantities because of preferential freight rates extended by the Denver and Rio Grande. The

Trinidad Coal and Coking Company, a subsidiary of the Atchison, Topeka and Santa Fe, supplied that line for its own consumption and for shipments to consumers in eastern Colorado and western Kansas.[36] That enterprise was not a serious rival for the major Colorado markets until after 1880, when additional railroads linked the Purgatoire Valley and the state's principal population centers.

The coal and town company's favored treatment from the Denver and Rio Grande had been formalized in a contract of May 20, 1876. In return for financial assistance in building an extension to El Moro and to fulfill earlier agreements with Stephenson and Jackson, trustees, the railway extended for a period of 20 years reduced freight tariffs on the transportation of coal and coke from the ancillary enterprise's lands in Huerfano and Las Animas counties. The rate was 20 percent less than the charge for a similar service to other shippers of the same commodities from that district.[37]

El Moro coal, because of its peculiar metallurgical qualities, was widely used throughout Colorado. The Denver Rolling Mill Company consumed up to 25 tons daily in its heating furnaces. The Boston and Colorado Smelting Company's plant at Argo, a suburb of Denver, mixed equal quantities of El Moro and Canon fuels in the reduction of precious metals. The Denver Gas Company employed the Trinidad-area coal in manufacturing illuminating gas. In addition, blacksmiths and foundrymen preferred that fuel in their forges and furnaces.[38]

Prior to January 23, 1880, the Southern Colorado Coal and Town Company sold about 78,000 tons of coal, three-quarters of which came from the Engleville Mine.[39] Almost one-half of the fuel produced in Las Animas County was shipped to the El Moro ovens for the manufacture of coke, which was by far the corporation's most profitable endeavor.[40]

In 1876, the company inaugurated experiments with the manufacture of coke. That year, Captain John Anderson erected six beehive ovens one-half mile south of El Moro to test the coking quality of coal from the mine at Engleville. When the results proved satisfactory, Charles Innes, a Trinidad contractor, was hired to build eight more units of the same design a mile south of the town where there was ample water and space for future expansion of the plant.[41] From December 1877 to October 1878, 14 ovens were in almost continuous operation. In the latter year, the original ovens were dismantled, and at the newer site, 62 units were constructed, financed in part by a loan of $10,000 from James B. Grant & Company, a Leadville smelter operator.[42] Thereafter, coke-making was confined to the one location at El Moro.

Resembling an old-fashioned beehive, an oven was constructed of fire brick with a circular base, a spherical dome at the top of which was a "funnel" through which coal was admitted and smoke and gases escaped during the coking process. At the front of each unit was an arched door for drawing or pulling the finished product.[43]

Slack and screenings from the El Moro Mine were sent to the ovens where each unit was charged while red hot with the fine-textured coal. The pile was leveled, and then the door was sealed with the exception of a small hole to admit air which swept over the fuel and out the funnel. Volatile matter distilled from the coal by the temperature of the oven ignited in the dome, the heat of combustion radiating downward, causing a swelling and fusing action. This created a desirable cellular structure in the end product. After 48 hours, the door was removed and the coke carefully watered to quench the fire without appreciably cooling the walls and base. After the oven had been allowed to "steam off," the coke was drawn onto the wharve or platform by one man using a "ravel" or scraper. Laborers using forks loaded the material on railway cars. The tined instrument was employed rather than a shovel in order to separate large pieces from ash and the fine particles of coke called "braize." Each unit produced about three tons of coke in a "burn," and the yield was approximately two-thirds of the original charge.[44]

The Southern Colorado Coal and Town Company's coke found a ready market in the West. The initial shipment, in February 1878, was to the Harrison Reduction Works in Leadville. That year, Pueblo's smelters consumed as much as 50 tons per day, and the El Moro product was also used in furnaces and foundries at Lake City, Golden, Boulder, and Denver. A total of 860 tons was ordered by smelters in Utah, which, until then, had relied on Connellsville coke from Pennsylvania.[45]

El Moro coke compared favorably with the eastern product except in quantity of ash. In an effort to correct this deficiency, the coal and town company in 1878 sent a carload of El Moro coal to the Carnegie enterprise in Pittsburgh where tests indicated that by crushing and washing the fuel before charging it in ovens, a superior grade of coke could be manufactured in southern Colorado. In this way, the ash content could be reduced from 18 to about 10 percent.[46]

In order to finance improvements at the El Moro Mine and ovens, the Southern Colorado Coal and Town Company, in 1878, mortgaged 10,000 acres as security for $150,000 in 10-year bonds bearing 10 percent interest annually. Two-thirds of that issue was immediately subscribed, and the balance was withheld from the market until needed for further expansion of business activities.[47]

Proceeds of the bonds were used to make improvements that more than doubled the output of the mine and ovens. In September 1878, a patented Stutz crusher and washer was erected at El Moro.[48] By that date, work had commenced on 180 new ovens. At Engleville, new entries were driven and the output of the mine increased to about 1,500 tons per day to supply a larger commercial market as well as the additional coke ovens.[49]

The expansion program occurred at a time when the company's titles to its lands in Huerfano and Las Animas counties were challenged by the federal government. On September 5, 1878, Westbrook S. Decker, United States Attorney for Colorado, filed in the Circuit Court, Denver, a suit charging that the enterprise and W. G. Riffenberg had committed fraud some five years earlier in acquiring 153 acres from the public domain near Trinidad.[50]

That legal action grew out of long-standing complaints of the manner in which land had been acquired from the federal government. In 1873, E. J. Hubbard, a Trinidad lawyer and horticulturalist, sent to Willis Drummond, Commissioner of the General Land Office, a letter advising that gigantic frauds had been committed in southern Colorado. Coal lands, the writer insisted, had been entered as agricultural claims by straw men who subsequently conveyed their rights to persons who pretended to be innocent of any wrongs. The Register and Receiver of the District Land Office in Pueblo were implicated in the conspiracies.[51]

Acting on this information, Drummond ordered the withdrawal of all lands in the Trinidad region until their true character had been determined. Also, he advised Hubbard to submit evidence of frauds to a grand jury and promised the support of the General Land Office in securing convictions of all guilty parties.[52]

In the autumn of 1874, a grand jury was empaneled in Pueblo to investigate charges that corruption existed in the District Land Office. On December 10, that body reported to Judge James B. Belford that "great frauds" had been committed by the employment of fictitious entrymen and witnesses on applications for quarter sections of land from the public domain.[53] That document contained a recommendation that the federal government institute suits to annul all patents that had been acquired in an illegal manner. However, no action was taken pending further investigations by appropriate agencies.

Three years later, Albert G. Stark and Abraham Bloomfield, residents of Las Animas County, petitioned the Department of Justice for permission to initiate in the name of the federal government a suit to cancel a patent allegedly obtained through fraud to a tract of land adjacent to the townsite of Starkville. Attorney General Alphonso Taft granted the request and instructed the United States Attorney in Denver to file legal action provided the petitioners agreed to protect the government against any costs.[54]

When that condition had been observed, W. S. Decker petitioned the federal court for an order cancelling the patent in question. He charged that Stark had taken possession of a tract, then part of the public domain near

Trinidad, on August 1, 1872, but when he subsequently filed a claim in the District Land Office in Pueblo, Register Irving W. Stanton and Recorder Charles W. Cook informed him that it could not be processed because the land was part of a Mexican Grant. However, one year later, the same officials approved preemption of the property in the name of James V. Snell, who was supposed to have resided there since 1872. The patent was conveyed to William S. Jackson, trustee, who later transferred it to the Southern Colorado Coal and Town Company. Two years later, the suit was dismissed without prejudice.[55]

On January 31, 1881, Decker instituted a similar but separate action. He charged that the Southern Colorado Coal and Town Company had conspired to bring about the dismissal of the earlier suit by purchasing Stark's interest in the Trinidad Coal and Mining Company, to which he had transferred all rights in the property at issue, at a price equal to five times their real value. That acquisition had been assigned to W. G. Riffenberg for the benefit of the corporation.[56]

Although the original suit had been initiated by the federal government in behalf of Stark and Bloomfield, the real instigator of the legal action was probably the Trinidad Coal and Coking Company, a subsidiary of the Santa Fe Railway. That enterprise hoped to bring about the cancellation of the patent by which a rival firm controlled one of the choicest coal mines in the Trinidad area.[57]

While involved in the Stark-Bloomfield suit, Decker was approached by other parties who, because of the government's action against the coal and town company, asked for the privilege of commencing suits to force the cancellation of patents on other lands owned by the corporation in the Trinidad area. This prompted a full investigation into the extent of frauds, leading Decker to conclude that they were indeed widespread.

In February 1879, Decker singled out six more patents for legal action. Two in Fremont County comprised the lands that had been acquired by Thomas T. Slocum and Joseph T. Musser on Coal Creek. The other four were located along the coal outcrop southeast of Trinidad. Decker charged that the Slocum-Musser lands had been claimed improperly as agricultural tracts. In securing the four claims near Trinidad, preemptors, according to Drummond, who served at this time as a special agent for the General Land Office, had not resided on the lands and were unknown in the community. Furthermore, it was commonly believed that the entries had been made by employees of the Denver and Rio Grande Railway "without settlement or legal intent."[58] All of the claims had been purchased by Jackson, trustee for the Trinidad Pool.[59]

One of the disputed claims, a quarter section in Section 29, Township 33, had been preempted in 1873 by Edward Soliss and conveyed to W. G. Riffenberg. A coal bank on the land had supplied local trade until, four years later, the Southern Colorado Coal and Town Company opened its Engleville mine on the tract. Drummond insisted that Soliss, who was unknown except to Riffenberg, was a fictitious person and that the property had been acquired from the public domain by fraud.[60]

Following a personal examination of the coal lands in southern Colorado in the spring of 1879, Decker reaffirmed his earlier view that wholesale frauds had been committed in Las Animas County. Thousands of acres, he stated, had been obtained illegally by Jackson for the Southern Colorado Coal and Town Company. The federal attorney recommended legal action to recover the lands; he believed that it was possible to prove that the men who had organized the coal and town company had engaged in a conspiracy to defraud the government of an immense amount of land.[61]

Evidence against the alleged conspirators, according to Decker, could be obtained from Dr. Michael Beshoar, who was at one time a party to the fraud. He agreed to act as a witness for the government in return for protection for some 3,000 acres of coal land that he claimed in the locality. Secretary of the Interior Carl Shurz, on reviewing the evidence, convinced Attorney General Charles Devens that the Department of Justice should not accede to Beshoar's demand. Schurz was confident that the government could support charges of fraud against the alleged conspirators without validating titles to land which he believed Beshoar had gained by fraud.[62]

On March 18, 1881, the United States Attorney filed in the Circuit Court at Denver a single suit that embraced sixty-one tracts in Huerfano and Las Animas Counties, for which titles were held by the Southern Colorado Coal and Town Company. By that date, however, the enterprise had been absorbed by the Colorado Coal and Iron Company. The successor firm had to engage in a prolonged legal contest to retain approximately 10,000 acres, including some of its best coal lands in southern Colorado.[63]

General Palmer and his associates merged the Southern Colorado Coal and Town, Central Colorado Improvement, and Colorado Coal and Steel Works companies to form the Colorado Coal and Iron Company for the purpose of bringing under a single management the raw materials required for the manufacture of iron and steel at Pueblo. Whether intentional or not on the part of the organizers, the new enterprise could claim that it was an innocent purchaser of the disputed fuel lands in southern Colorado; therefore, its titles were valid even if fraud had been committed at some earlier date in securing the properties from the public domain.

In nearly five years of corporate existence, the Southern Colorado Coal and Town Company exerted an enormous influence on Colorado and the Rocky Mountain region. Coal and coke from Las Animas County stimulated the growth of metallurgical industries, especially the processing of nonferrous ores. That business, after 1876, grew rapidly as El Moro coke supplanted charcoal as the basic fuel. At the same time, reduction furnaces and refineries, previously concentrated in Leadville and other mountain towns, were erected in Denver and Pueblo, which became the principal center for the treatment of precious metals.

The coal and town company did not pay dividends to stockholders. Net earnings of the two mines, one of which operated at a loss, and the El Moro ovens exceeded $35,000, from which corporate salaries, general expenses, taxes, and interest on debts were paid, leaving a surplus of less than $9,000.[64] This small return for a million-dollar enterprise may have influenced stockholders to approve, on December 13, 1879, the proposed consolidation with the Central Colorado Improvement and the Colorado Coal and Steel Works companies.[65] The new scheme to consume wholly owned raw materials in the manufacture and fabrication of iron and steel held out the promise of large profits to investors who had not benefited until that time from Palmer's real estate and industrial ventures.

Walsenburg Area Camps

Mine buildings at Rouse, ca. 1915.

Rouse mine was opened by the Colorado Fuel and Iron Co. in 1899 near the site of what had been the Colorado Coal and Iron Co.'s Santa Clara slope, operational from 1889 to 1891. Old Rouse, opened in 1888, was abandoned in 1899 because of the high cost of pumping excessive water from underground workings. New Rouse, located 12 miles south of Walsenburg, was abandoned in 1920 because of flooding. The slope produced 4.1 million tons of coal.

Mine office at Rouse, May, 1915. From left to right, Gordon Scott, Dad Wright, and E. Erskine.

The most pretentious structure in any coal mining camp was the superintendent's residence. Standing on the steps of his home at Rouse ca. 1915 are superintendent W.G. Deck, and two lesser officials, E. Erskine and a man named Elliott.

Exterior of boarding house for coal miners at Rouse, ca. 1915.

Miners at mealtime in boarding house at Rouse, ca. 1915. The mode of dress suggests this was a formal occasion, perhaps Sunday dinner.

Park and playground at Rouse, Oct. 1915.

Osgood School at Rouse, ca. 1915. The building, named for the first president of the Colorado Fuel and Iron Co., was of the standard design built in numerous camps by the firm's sociological department in the first decade of the present century.

U.S. Post Office at Rouse, May, 1915.

4

Colorado's First Rolling Mill

The Panic of 1873 and ensuing depression eliminated at least temporarily the railway promoters' plans for manufacturing and fabricating iron in southern Colorado. Palmer and his associates had hoped that the Central Colorado Improvement and the Southern Colorado Coal and Town companies could erect blast furnaces and rolling mills at either Canon City or South Pueblo. These men believed that rails, which the Denver and Rio Grande required in large quantities for construction and repairs, could be made more cheaply in Colorado than imported from eastern states or abroad. In addition, the ancillary companies wanted to exploit for their own profit, and as a source of traffic for the narrow gauge line, vast deposits of coking coal, limestone, and iron ore in Fremont and Las Animas counties.

Following the collapse of their own schemes, officials of the Denver and Rio Grande encouraged others to erect facilities for producing iron products in or near South Pueblo. The railway joined with local businessmen in encouraging William A. McKinlay, part-owner of the Licking Iron Company of Shawnee, Ohio, to make in August 1877 a thorough investigation of the iron market and the availability of resources in Colorado. He concluded that while the seven foundries operating in the state consumed only ten tons of iron weekly, with the erection of a rolling mill the demand would increase to about 1,000 tons annually, sufficient to justify the construction of a small blast furnace.[1] Furthermore, coke from the ovens at El Moro was ideal for smelting iron, and the ores that abounded in Fremont and Las Animas counties were suited to the production of pig and mill iron.

Although that scheme failed to materialize, William Faux of Danville, Pennsylvania, who had accompanied McKinlay to Colorado, opened negotiations with Pueblo businessmen in an effort to solicit financial inducements for relocating his iron rolling mill in the West. At a public meeting on the night of August 28, local residents learned from Charles B. Lamborn, vice president of the Central Colorado Improvement Company, that "Pennsylvania parties" were prepared to erect in the city a foundry, machine shop, and rolling mill, the latter with a capacity of from 30 to 40 tons of railroad iron per day.[2]

The iron-maker, Lamborn recounted, asked for liberal amounts of land and money. He wanted a five-acre site near railroad facilities and supplies of water, real estate worth $40,000, and cash subscriptions to the capital stock of his company.[3]

The editor of the *Pueblo Chieftain,* convinced that the city's destiny depended upon industry, urged readers to contribute whatever was necessary to secure the mill. Its value to the community was incalculable because it would attract people and money, and focus attention on Pueblo as a manufacturing center, inducing other enterprises to locate there.[4]

Negotiations between Faux and Pueblo businessmen dragged for nearly a month. On September 21, he finally agreed to move his plant if given land worth $20,000, plus subscriptions of $12,000 to the stock of the Pueblo Iron Company. Before the close of that day, the officials of the Denver and Rio Grande Railway and local men had pledged the requisite real estate and money.[5]

The narrow gauge line, on Septmeber 22, formally established tariffs on freight for the Pueblo Iron Company. By terms of that agreement, coal, coke, and iron were to be transported at low rates for a period of five years.[6] In addition, the enterprise was assured unlimited supplies of raw materials at reasonable prices. Canon, Cucharas, and El Moro coals were to cost from $3.00 to $3.50 per ton, iron ore was to be mined from lands of the Central Colorado Improvement and Southern Colorado Coal and Town companies at a royalty of only five cents per ton, and fire clay was to be excavated without charge. In return, the railway expected to purchase rerolled rails and castings for its own use at fair prices. The proprietor of the rolling mill guaranteed that the cost of these products, over a period of five years, would not exceed schedules for similar materials in Chicago, St. Louis, Kansas City, and Springfield, Illinois.

The articles of incorporation for the Pueblo Iron Company, filed under Colorado law on September 24, 1877, authorized the enterprise to construct and operate a foundry, machine shop, and rolling mill for the purpose of manufacturing railroad and merchant iron, castings, and other products.[7] The principal subscribers to the capital stock of $150,000, other than Faux who received shares equivalent to the value of the plant, were Mahlon D. and John A. Thatcher, O.H.P. Baxter, William Moore, J. V. Andrews, James N. Carlile, Moses Hallett, Henry C. Thatcher, George H. Hobson, M. H. Filet, Charles Gast, and Charles Henchel and Company.[8] Directors for the first year, in addition to Faux, were M. D. Thatcher, Baxter, Andrews, and Moore.

The new industry was located on Santa Fe Avenue between the old and new channels of the Arkansas River. Two large buildings were erected to house several carloads of machinery and equipment that reached Pueblo in mid-December. Sixteen mechanics who accompanied Faux from Danville installed eight boilers, three engines, cupolas, blowers, pumps, rail and bar mills, saws, and shop equipment. The cost of relocating the plant from Pennsylvania to Colorado came to $26,690, about one-fifth of which was transportation charges.[9] When rebuilt, the plant had a rated daily capacity of from 30 to 40 tons of T-rails and from 8 to 10 tons of merchant iron.[10]

Convinced that it would be foolish, even ruinous, to import pig iron when materials for its manufacture abounded in southern Colorado, the promoters of the rolling mill sought to bring to Pueblo an operator of a blast furnace. In the fall of 1877, the rumor circulated throughout the city that a plant was to be erected and that the Denver and Rio Grande was preparing to extend its Arkansas Valley branch from Canon City to iron-ore deposits on Grape Creek.[11] William Faux and Captain James Rice, former mayor of Pueblo, went East in January 1878, to talk to experienced ironmakers who had expressed an interest in relocating in the West.[12] In the next two months, at least three Ohio furnacemen visited Pueblo to examine the region's resources for the manufacture of iron. When none chose to move to the city, Faux decided to start up his rolling mill on old rails and scrap metal, which imposed severe limitations on the variety and volume of its output.

General Palmer negotiated in February a contract with the Pueblo Iron Company for the Denver and Rio Grande's requirements of iron rails. The rolling mill agreed to supply the narrow gauge line at a price of $25 per ton for 30-pound track, and to accept two and one-half tons of old rails in exchange for one of the rerolled product. If Faux had to purchase iron elsewhere, the railway agreed to pay the actual costs plus a fixed rate for processing, with the privilege of remitting one-half in 90 days, with notes for the balance due over a period of ten months at an interest of 10 percent per annum.[13] Faux was to produce as much as 30 tons of T-rails daily to fill the carrier's needs for construction and renewals. He also agreed to proceed as rapidly as possible with rerolling of 500 tons of old rails on hand.

Production commenced at the Pueblo plant in early March. The heating furnace was charged with iron on the fourth of the month and rails were rolled the following day.[14] Output was severely limited for at least a week because of frequent breakdowns and stoppages for adjustments of the machinery.

On March 18, approximately one week before the first rails were shipped from Pueblo, Faux modified his contract with the Denver and Rio Grande. In settlement of monthly purchases, the mill proprietor agreed to accept one-third cash and the balance, secured by notes at 15 percent interest, over a period of a year. At the same time, the railway ordered another 3,000 tons of rails at a cost of $13,500 to complete its Alamosa extension.[15]

Rails were rerolled from discarded wrought-iron track. Fabrication commenced with the cutting of a rail section into three or four pieces. These were bound with wire to form a compact mass, or fagot, which was subjected to a welding heat in the furnace. At the desired temperature the material was repeatedly passed through the 10-inch mill, the shape and pressure of the rolls transforming the iron into a fagot bar or flat. That section was cut into three or four pieces, which were bound, heated and rolled into a billet. After reheating, the billet, on passing through the roughing and finishing rolls of the 16-inch mill, emerged as a T-rail. The final product was cut to approximate length while hot and transferred to a cooling stand. If an exact length were required, the section had to be recut with a "cold saw."[16]

The production of rails was interrupted in early April when heaters and rollers walked off their jobs. The strikers complained that the wage scale was inadequate. Each man, they insisted, had expended from $40 to $50 in moving to Colorado, and more if he brought a family, only to find that the cost of living was higher than in the East. However, the Pueblo mill used a piece-work rate of compensation that prevailed in Springfield, Illinois.[17] An adjustment was made. By mid-May, the plant was running a double "turn" in manufacturing "heavy iron" for the Atchison, Topeka and Santa Fe.[18]

Over a period of three months, the Pueblo Iron Works shipped almost 900 tons of rails to the Denver and Rio Grande alone. Only a small proportion of that amount had been rerolled from the railway's old track; the balance was fabricated from iron purchased at Emporia, Kansas, from the Missouri, Kansas and Texas Railway by the Colorado-based carrier. Faux charged $28,608.77 for processing, a figure that was disputed by the purchaser on the grounds that he had not deducted 15 percent for "wasteage."[19] This controversy, the outcome of which is unknown, may have influenced the rail-maker to seek another location for his mill.

In the spring of 1878, as another surge of railroad construction occurred in the West, numerous communities, among them Denver, sought to attract furnaces and rolling mills as local industries. That year, ex-Governor John Evans and other businessmen corresponded with eastern iron manufacturers and fabricators who were willing to consider relocating in Colorado's capital city.[20]

In May of that year, John W. Smith, pioneer Denver miller, and George Tritch, local hardware merchant, received from an "Eastern manufacturer, a statement of the conditions under which he was willing to relocate his rolling mill in Denver. He demanded a plant site free of charge, contributions of not less than $5,000 in cash, and subscriptions totaling $10,000 to the stock of a new enterprise, the Denver Rolling Mill Company, to be incorporated with an authorized capital of $100,000. Two of the firm's directors were to be residents of the city. Buildings were to be erected and the machinery installed, ready for operation within 90 days after arrangements had been made for financial assistance.[21]

Presumably, the iron-manufacturer approached Smith and Tritch because they were prominently identified with Governor Evans' Denver South Park and Pacific line then under construction to Leadville. Along with Frederick J. Ebert and others, they had organized in 1877 the Denver and South Park Construction and Land Company, of which Smith was the president.[22] Under his direction, the narrow gauge line was completed in 1880 to the silver camp.

The easterner's proposition, according to Tritch, was Denver's golden opportunity to secure "a genuine good thing at a reasonable cost." The mill owner, an experienced ironmaker, recognized that Colorado's capital city was the right place for his enterprise; all he required was assistance to get started.[23]

The Denver business community threw its support behind the movement for the local iron industry. The community's Board of Trade adopted on May 29 a resolution favoring acceptance of the proposition and appointed five men, including Smith and Tritch, to solicit donations of land and money to bring the rolling mill to the city.[24] The editor of the *Daily Times* urged support for the fund-raising drive, pointing to the advantages that Pueblo had reportedly derived from its iron works. At that time, he did not know that the unnamed applicant for financial assistance was William Faux and that his plant was even then in operation in the state of Colorado.[25]

By mid-June, rumors of Faux's negotiations circulated in Pueblo. When confronted with the reports, he acknowledged their accuracy and insisted that the rolling mill was to be relocated in Denver.[26]

The impending move touched off a spirited debate between newspapers in Pueblo and Denver. The editor of the *Chieftain* accused Faux of operating a "racket" in which he set up his dilapidated mill in a city, collected prearranged subsidies, and then opened negotiations with another community that was anxious to foster the growth of local manufactures.[27] The Pueblo Board of Trade, in an effort to apprise Denver of the true character of the man with whom it was dealing, charged Faux with systematic misrepresentation, duplicity, and fraud "in the manner in which he attempted to rob our city of its liberality."[28]

The *Denver Daily Times* rushed to Faux's defense. It insisted that his decision to relocate the mill was based on the knowledge that a Pueblo-based iron works could not compete with a similar plant in the state's capital city. By moving to Denver, Faux would not only forestall competition but at the same time gain access to a larger market. Only two railroads served the southern city, while six terminated in Denver.[29]

In his own defense Faux insisted that the residents of Pueblo had not fulfilled the terms of their contract with him. At a cost of $20,000, he had removed his mill to that city in return for promises of $25,000 in subscriptions plus valuable real estate. Only "something over $9,000" and the site for the plant had been donated.[30] The latter was deeded with a proviso that for a period of five years it was to revert to the Denver and Rio Grande if the mill were moved or destroyed by fire and not rebuilt.

Charges and counter-charges continued for several weeks as Faux and his Pueblo benefactors aired their complaints in the daily press. Col. M. H. Fitch, president of the Stockgrowers' National Bank, asserted that some real estate had been withheld becaüse the iron-maker had not given bond for the maintenance of the mill for five years, nor had he established a foundry and machine shop as required by his contract.[31] Faux insisted that the newly opened Pueblo Foundry and Machine Shop had fulfilled the latter obligation. John Fiddler and I. C. Van Fossen, partners in the business, acknowledged that they had purchased machines and equipment from the Pennsylvanian but denied that they were in any way associated with him or a part of his agreement with local investors.[32]

At the height of the controversy, the minority stockholders of the Pueblo Iron Company petitioned the county court for an injunction to prevent relocation of the rolling mill. Attorney Charles E. Gast initiated the legal action in the name of the local directors.[33] When an order was issued restraining the officers of the enterprise, Faux and his Denver associates had no alternative but to seek a financial settlement with the Pueblo investors.

Arrangements were completed that summer for removal of the mill. W. G. Sprague, a Denver real estate broker, was commissioned to purchase the Pueblo-owned stock in the iron company. Initially he offered $.65 on a dollar.[34] Since there was no rush to sell at that price, he must have paid more, perhaps even full value for the shares.

Negotiations were momentarily marred by a threatened strike of the mill's employees. In early August, some of Faux's checks for overdue wages were refused by a Pueblo bank because of insufficient funds in the company's account.[35] A crisis was averted when the Denver investors promised immediate and full payment of all monies due the workmen.

When it became obvious that the mill was to be transferred from Pueblo, the *Chieftain* fired a parting shot at the enterprise and its owner. The community, according to the editor, "would have been better off . . . had it never been blessed with the presence of Mr. Faux and his twenty-seven cars of iron."[36] The eastern entrepreneur had held up a glittering bait to residents of Pueblo who, in their desire for local manufactures, forgot the homely proverb: "Treat every man as a rogue until you know him to be honest."[37] No attempt had been made to ascertain the truthfulness of Faux's statements, no committee had been sent to Pennsylvania to examine the rolling mill or to determine its real value, and no inquiry had been made about the owner's business practices. Accepting Faux's word as gospel, local investors had surrendered their hard-earned money.

It was an unfortunate experience that raised suspicion about all enterprises that might be attracted to Pueblo, and that attitude could paralyze progress in manufacturing. Only time, the newspaper continued, "will dispel the soreness produced by the present failure and then honest men with honest schemes can again obtain solid aid."[38] However, in the future, investors will scrutinize "the character, standing and pecuniary resources of the applicant" before committing capital to a new industry. Meanwhile, Pueblo had lost its iron works.

The Denver Rolling Mill Company was incorporated August 29, 1878, by William Faux, his son William J. Faux, John Lister, John L. Fields, and Frederick J. Ebert. The incorporators were the directors for the first year.[39] The enterprise had an authorized capital of $150,000.

That month, after pledges of money and land amply met all conditions prescribed by Faux, the Pueblo mill was dismantled and transported to a six-acre site on the west side of the fair grounds and adjacent to the Kansas Pacific's tracks at Thirty-second and Wazee streets in North Denver. The task of reassembling the plant had progressed sufficiently by mid-September that Faux requested the initial installment of cash and subscriptions to stock.[40]

In early November, the mill resumed the manufacture of iron rails. A fire had been started in the heating furnace on October 31, and rails came off the 16-inch roll train ten days later. The initial product was set aside by Capt. Ebert as a symbolic "mile-stone in Denver's commercial progress."[41] The next day, work commenced on a 500-ton order for the Denver South Park and Pacific. Production continued at the rate of from 20 to 30 tons daily.

Colorado's expanding railroad network seemed to assure the success of the rolling mill. On completion of the initial order, Faux anticipated another of equal size from the Denver South Park and Pacific. He also expected the Union Pacific, Kansas Pacific, Atchison, Topeka and Santa Fe, Colorado Central, and Denver and Rio Grande to make heavy demands on the capacity of the plant.[42]

In spite of seemingly bright prospects, under Faux's management the enterprise lasted only a few months. Production was frequently interrupted by accidents and suffered from a conspicuous absence of demand for rails and other iron products. Chronically short of money, Faux fell in arrears in wages to his employees and ignored his creditors. In mid-December, operations were suspended for several days. Later that month, the Colorado Iron Company filed a lien on the Denver Rolling Mill Company's land and machinery — following which, the plant closed "for want of work."[43]

The company was reorganized in 1879. Available evidence, although inconclusive, suggests that Faux surrendered management of the enterprise to Sprague and Ebert. By 1880 the mill was running at full capacity, turning out rails and bar iron for local blacksmiths and manufacturers, and the owners seriously considered new buildings and machinery to enlarge the output.[44]

For a short time, William Faux and his son actively engaged in mining enterprises in Lake and Summit counties. The older man was one of the incorporators of the Presbyterian Gold and Silver Mining Company, while the younger one was an officer of the Wyoming Mining, Smelting and Milling Company, and an incorporator of the Colorado and Leadville Railroad Company.[45] After 1881, both men vanished from the Colorado scene.

In the period from 1878 to 1880, Faux's rolling mill was the only manufacturer of iron rails in Colorado. In that time, the output amounted to 8,317 tons, or 2,772 per year, all of which were purchased by the Denver and Rio Grande, the Santa Fe, and Denver South Park and Pacific.[46] Since the plant had a capacity of about 10,000 tons per year, the small volume suggests there was little demand for rerolled iron rails. The company, certainly in Denver if not in Pueblo, also manufactured a large amount of merchant iron.[47]

The removal of Faux's mill from Pueblo had prompted General Palmer, Dr. Bell, and C. B. Lamborn to revive plans for an iron works of their own. By 1879, when the increasing tempo of business activity justified a new venture, they recognized that because of heavier rolling stock and higher operating speeds, railroads had to turn to the more durable steel track. Therefore, they envisioned an integrated plant capable of producing from its own raw materials pig iron for manufacture into steel, to be fabricated into rails and other products for which there was a market in the West. Because of the continuing demand, the new facility was also to produce iron products, including light rails, merchant iron, cut nails, and spikes.

On October 22, 1879, General Palmer, C. B. Lamborn, and Lyman K. Bass organized the Colorado Coal and Steel Works Company. Articles of incorporation filed three days later in Denver authorized the enterprise, among other things, to "erect and operate furnaces and manufactories of iron and steel, and to sell products of same, and to build and operate coke works."[48] The capital stock was fixed by the charter at $5,250,000. The board of trustees for the first year were the incorporators, who elected Palmer as president and Dr. Bell as secretary-treasurer.[49]

While the charter of the Colorado Coal and Steel Works Company suggests that it was formed to manufacture iron and steel products, in fact it was a paper corporation that served only to channel profits from the formation of an integrated metallurgical works into the pockets of a few promoters. There are no records of the enterprise's stockholders, other than incorporators and officers, or of any activities in which it engaged except its merger with the Central Colorado Improvement and the Southern Colorado Coal and Town companies in January 1880 to form the Colorado Coal and Iron Company.[50]

That new corporation, while engaged in construction of an iron and steel plant at South Pueblo, acquired by purchase in August 1880 the property of the Denver Rolling Mill Company. The initial cost was almost $43,000, and an additional $52,000 was invested in reconditioning and modernizing the plant. New boilers, a warehouse, a sheet-metal structure to enclose the 10-inch mill, additional shears and saws, and equipment for making a wider range of iron sizes and shapes advanced the mill's capacity to 40 tons daily. The new owner also planned to install a puddling furnace for handling pig iron in order to avoid complete dependence on old rails and scrap metal.[51]

The Denver Rolling Mill, as the plant was commonly known, remained in operation until the end of 1884. By that time, the Colorado Coal and Iron Company found it more economical to fabricate light rails and merchant iron at Pueblo where rolling mills recently purchased from the Ogden Iron Works had ready access to pig as well as scrap iron.[52] However, production resumed in Denver during the next two years whenever there was an unusually heavy demand for light rails or merchant iron.[53] The plant was finally closed in 1886. Three years later, the buildings were dismantled, and all machinery and equipment was transferred to the Pueblo plant. The 10- and 16-inch roll-trains, together with heating furnaces, cupolas, and related appurtenances were added to the merchant iron department where they continued in operation for many years.[54]

The Colorado Coal and Iron Company's decision to consolidate all iron-rolling operations was the final blow to Denver's dream, initially kindled by William Faux's mill, of becoming the center of heavy industry in the Rocky Mountain region.[55] Thereafter, Pueblo was the focal-point of iron and steel manufacturing in the West.

5

The Colorado Coal and Iron Company, 1880-1884

At Colorado Springs, on December 13, 1879, stockholders of three enterprises ratified a plan of consolidation that created the Colorado Coal and Iron Company with an authorized capital of $10,000,000. The stock of the new venture was to be exchanged for that of the older corporations: one share for one of the Central Colorado Improvement Company; two and one-half for one of the Southern Colorado Coal and Town Company; and five for seven of the Colorado Coal and Steel Works Company.[1] Each of the merging firms was entitled to name one member of the new board of directors. Those three men, in turn, were to elect two others, and the fully constituted body was to organize the successor, at which time the consolidating units were to surrender all properties, seals, transfer books, and other records.[2]

Representatives of the consolidating companies met on December 16, 1879, to fill vacancies on the board of directors. Lyman K. Bass, C. B. Lamborn, and Robert F. Weitbrec selected for those positions General Palmer and Charles F. Woerishoffer, a New York financier. At that time, the board elected officers for the first year. Palmer was named president, Lamborn vice-president, and Walter B. Gaskill secretary and treasurer. The certificate of consolidation and articles of incorporation were filed with the Colorado Secretary of State on January 23, 1880, at which time the Colorado Coal and Iron Company assumed all functions of the predecessor firms.[3]

The company's varied activities were placed under the direction of Appleton H. Danforth, who was named general manager on January 23, 1880. That month he was also appointed secretary of the corporation, a post he held for about a year before final preparations for the manufacture of iron and steel required that he devote his time to the operating departments. He also served as a vice-president of the enterprise for several years.[4]

Subordinate officers were men of broad experience in their respective lines of endeavor. A. G. Haumann, a consulting engineer from Pennsylvania, was placed in charge of iron mines. James R. Cameron, formerly general manager of the Killington Coal Company, Osceola, Pennsylvania, was given the responsibility for operations of coal mines and coke ovens. Caspar Constable, an experienced iron-master, was named superintendent of the steel works. Cameron died in October 1881 and was succeeded by John Cameron, who was appointed supervisor of all mining activities when Haumann resigned three years later.[5]

The board of directors, on February 1, 1880, authorized a mortgage of $3,500,000 on the company's properties to retire the bonded indebtedness of predecessor enterprises and to provide capital for new business activities. After all obligations had been met, there remained $775,000 in bonds, augmented by capital stock worth $262,500, to underwrite operations at South Pueblo and elsewhere in southern Colorado.[6]

The company planned to engage in four activities, all but one of which had been inaugurated by the predecessor enterprises. These included the improvement and sale of town lots and agricultural lands, the production of coal, and the manufacture of coke. In addition, Palmer and associates proposed to establish a plant to manufacture iron and steel products.[7]

These functions were to be sustained by nearly 100,000 acres of land acquired from predecessor enterprises. That amount included 13,571 acres of coal deposits in Fremont, Huerfano, and Las Animas counties; 1,057 acres of iron ore on Grape Creek between Canon City and Silver Cliff; and 83,944 acres of agricultural lands in the valleys of the Arkansas, Cucharas, and Purgatoire rivers. In the latter category were the townsites of South Pueblo, Labran, Cucharas, and El Moro, plus a portion of Canon City.[8]

The coal properties, designated by the company the best of their class in Colorado, contained excellent domestic, steam, and metallurgical fuels. Nearly three-quarters of the holdings were near Trinidad; other large parcels were located near Walsenburg and in the Canon City district.[9]

The Colorado Coal and Iron Company was the largest coal producer in the southern field. From predecessors it had obtained the Engleville (El Moro) Mine southeast of Trinidad, the Walsen Mine at Walsenburg, and the Coal Creek Mine southeast of Canon City. In 1880, those properties accounted for nearly one-half of all the coal mined in Colorado.[10]

Until the Colorado Coal and Iron Company's iron and steel works commenced operations in 1881, the corporation's coke production was sold almost exclusively to smelters of precious metal ores. At that time, according to General Palmer, there were 74 plants in Colorado, most of them silver-lead stacks at Leadville. Each plant required about two and one-half tons of coke daily. In addition, Utah's smelters consumed another 100 tons per day. The market was supplied either from Pennsylvania or from Trinidad, where the El Moro ovens were the largest producers. The firm planned to add 180 units to the plant and eventually to raise the number to 400 in order to provide ample coke for the reduction of iron ore as well as an expanding smelter trade.[11]

Coke from Trinidad was to be combined with iron ore from Fremont County to produce foundry and pig iron. The company planned to mine magnetite from its lands on Grape Creek and, if necessary, to mix that product with carbonaceous ores that abounded in Las Animas County, as well as at other points throughout southern Colorado.

In order to secure economical transportation for coal, coke, ore, and finished metal products, R. H. Lamborn and Danforth negotiated on November 1, 1880, a contract with General Palmer as chief executive officer of the Denver and Rio Grande Railway. For a period of 20 years, the carrier agreed to transport all products of "mines, furnaces, smelting works, mills, and other establishments" belonging to the Colorado Coal and Iron Company "at rates never to exceed the best rates given to any other company, association or individual."[12]

South Pueblo was chosen as the site for the iron and steel works for several reasons. It was a transportation center where raw materials could be economically and conveniently brought together; the city's railroads afforded easy access to a vast regional market; and water was abundantly available. Furthermore, the company owned there a large acreage of townsite and other lands, the value of which, it was anticipated, would appreciate with the growth of heavy industry.[13]

On the mesa south of the river, beyond the town limits of South Pueblo, the company set aside 1,260 acres for the iron and steel works. The ground was surveyed in February 1880, and work commenced that month on a hot blast furnace, Bessemer converter, blooming and rail mill, foundry, and machine shops. Col. W. W. Taylor, brother of the late Bayard Taylor, noted poet, traveler and diplomat, was appointed superintendent of construction.[14]

On February 2, eight men began work on the foundation of the furnace. That force was under the supervision of James K. Dempsey, formerly of Pittsburgh; it included Michael Martin, foreman of the quarry; John McGuire and Charles McKenney, bricklayers; James Madden and George McKuen, quarrymen; and Charles Peterson and a man named Raper, laborers.[15] They erected a rough plank shelter as a boarding house and small office for the superintendent of construction.

As the labor force multiplied, clusters of makeshift homes blossomed around the construction area. West of the furnace site, the "rough and ready" town of Dodge City contained 75 "cabins," many built of scrap lumber, sheet metal, and canvas. Farther east along Routt Avenue, "Wichita" experienced a similar boom.[16]

Local residents named the settlement around the future steel works in honor of the superintendent of construction. "Taylorville" remained the popular name until 1881, when the company insisted that the proper designation for the area was "Steel Works." By that time, between 300 and 400 men were employed there, and the monthly payroll, to the delight of tradesmen on both sides of the river, amounted to from $7,000 to $8,000.[17] Later that year, the corporation began organizing a town which was called "Bessemer" and, through the South Pueblo Homestead and Investment Company, arranged for the erection of a large number of homes.[18] On completion of that construction program, the temporary dwellings were abandoned.

Even while the plant was under construction at Bessemer, the coal and iron company began manufacturing iron products in Denver. The Denver Rolling Mill, purchased in August 1880, was reconditioned, after which it produced rerolled rails and merchant iron for the Colorado market.[19] Operations declined in 1885 and ceased the following year because it was more economical to fabricate iron products at the steel works. Eventually, the buildings, rolling mills, other machinery, and equipment were transferred to the plant at Bessemer.[20]

The construction schedule at Bessemer gave top priority to the foundry and machine shop to enable them to manufacture parts for the steel works. Located along the northern boundary of the plant, these facilities were ready for production by February 1881. Under the direction of master mechanic L. J. Taylor, a force of 20 mechanics and moulders turned out roof beams, fire fronts for boilers, floor plates, charging larries for the furnace, ingot moulds for the converter, and a variety of castings for the coal and iron mines. In this way, the company saved up to 20 percent on the cost of many items.[21]

Most of the shop employees came to Bessemer from Pennsylvania. The chief mechanics were William Montgomery and George Henry. The moulders included Richard H. Allen, who was the foreman, William Longeron, E. C. Matson, Con Myers, Charles Ott, J. M. Schultz, and M. W. Smith.[22] Myers had been employed in the construction of the converter and puddling mill before joining the foundry crew. When he retired on January 1, 1928, he was the oldest employee in time of actual service, having spent 45 years at the steel works.[23]

On April 6, 1881, the Colorado Coal and Iron Company appointed Daniel N. Jones to the position of superintendent and chief engineer of the Bessemer Works. An experienced iron-maker, he had been superintendent of the Cambria Iron Company for several years before deciding to move to the West. Caspar Constable resigned that year to devote more time to business interests in the East.[24]

The blast furnace, first of two in the construction program, was "blown-in" September 5, 1881, in a formal ceremony that was attended by company officials and community leaders. Christened "Betsy" in honor of superintendent Jones' young daughter, it was designed to yield 80 tons of iron daily. The initial tap of 12 tons was made two days later. Thereafter, the unit remained in operation for more than two years before it had to be shut down for major repairs.[25] At first, slag was an important by-product, for it was used to ballast railway tracks and to fill depressions throughout the plant site.[26]

"Betsy," or Furnace No. 1, a second-hand unit the company had purchased in Ohio, was small even by the standards of that day. Standing 65 feet high with a maximum width of 17 feet, it was a brick shell with an inner lining of fire brick. Three Siemens-Cowper-Cochrane stoves generated hot air for the blast. A casting house contained sand moulds for making "pigs." The "storage house" was an open area traversed by a railway trestle from which carloads of coke, ore, and limestone were dumped on the ground. As needed, these materials were loaded into a charging lorry, hoisted to the top of the furnace, and dropped into the furnace stack.[27]

Under the watchful eye of superintendent Josuah Crowther, the blast furnace remained in operation day and night. Two shifts of 35 men each were directed by William J. Ripley and Joe Kendricks. In March 1882, poor health forced Crowther to return to his home in Pennsylvania. At that time, Ripley was placed in charge of the furnace and Kendricks was promoted to the position of foreman.[28]

Although the converter house, a brick structure with a tin roof, was completed in the final weeks of 1881, delays in the delivery of machinery and the time required to accumulate a large stock of pig iron caused a postponement in the start of steel-making until the following spring. The building contained two five-ton Bessemer vessels, cupolas for melting pig iron, blowing engines, overhead cranes, implements for crushing and grinding refractories, and an oven for drying converter bottoms and ladles.[29]

The production of steel commenced on April 12, 1882. In the early hours of the morning, pig iron, heated to a molten state in the cupola, was placed in one of the converters. At a signal from superintendent George J. Humbert, the blower opened valves that sent a flow of air through the metal. At the conclusion of the "blow" the material was recarburized with ferro-manganese, drawn off into a ladle, and poured into ingot moulds measuring 15 by 15 by 72 inches. After cooling, the moulds were stripped from the ingots, which were transported to the blooming mill and placed in a soaking pit.

The rolling mill was supervised by George Tatnall, assisted by John Potter, foreman of the blooming mill. Harry Fonlier was furnaceman, and Conn Finn and Bill Maddock were heaters. Among the rollers on the blooming and rail trains were John P. Allwine and David Lloyd. Jeff Fitzsimmons, John Russel, and Tom Roach were the rail straighteners.[30]

The rolling process commenced at eight o'clock in the morning. An ingot was removed from the soaking pit, placed on the blooming mill table, and passed through the rolls until it was reduced to an elongated rectangle six inches on each side. The ends were sheared, and the metal, in the form of a billet, was reheated before it entered the rail train. There it was shaped by the roughing and finishing rolls, emerging as a T-rail. The ends were sawed off and the 30-foot section was cambered before it was set aside to cool on the hot bed. From the treatment of pig iron in the converter to the final pass in the rolling mill, the manufacture of the first rail had consumed 11 hours.

Interested spectators to that historic event were two lifelong friends, Daniel N. Jones, superintendent of the Bessemer Works, and Captain William R. Jones, legendary iron-maker who had charge of Andrew Carnegie's Edgar Thomson Steel Works in Pittsburgh. Captain Jones, a man who could not resist an opportunity for a wager, bet his colleague that the initial rail would be defective.[31] The stake was a dinner for the working force at the city's best restaurant. The 30-pound section proved flawless. Subsequently, 200 men enjoyed a banquet at the Victoria Hotel as the guests of Bill Jones.

Although made on an order for the Denver and Rio Grande, the first rail was cut into small pieces and distributed as souvenirs to employees and local citizens. Two weeks passed before rails were again rolled at Bessemer. Shipments to the railway commenced at the end of April.[32]

In June, soon after Reese James succeeded Humbert as superintendent of the converter, the steel department went on double turn to keep pace with demand. That year, the rolling mill turned out about 1,000 tons of 56-pound patterns for the Chicago, Burlington and Quincy, plus 15,000 tons of T-rails for the Denver and Rio Grande's Silverton and Utah extensions.[33] One 30-pound section manufactured in September for the narrow gauge line remained in use some 50 years later in the Lark Mine near Brigham, Utah.[34]

Initially, the company had planned to establish a puddling mill in Denver to manufacture muck iron from scrap metal for the rolling mill. Because it was more economical to build in Bessemer, that plant was erected east of the blooming and rail mills.[35] The structure was completed in early 1882, at which time George Humbert and William Tatnall supervised the installation of a 20-inch muck train, cupolas, drills, presses, and other equipment required for the production of rolled wrought iron. That year, the mill turned out 1,411 tons of bars and plates, more than three-quarters of which was consumed in the manufacture of nails and spikes.[36]

The nail mill was located in the same building with the puddling mill. Twenty-seven machines had the capacity to produce all sizes of cut nails at the rate of about a million pounds a year. Under the direction of William Tatnall, nails were manufactured for the first time on September 13, 1882. The eight and ten-penny varieties were distributed as souvenirs. Before the close of the year, the output exceeded 160,000 pounds.[37]

Among the operators of the hand-fed machines were I. C. Van Fossen and his son Ike. The older man had moved to Pueblo from Wheeling, West Virginia, four years earlier to open a foundry and machine shop on Santa Fe Avenue near the river. There he also operated a small nail manufactory. When the Colorado Coal and Iron Company built its mill, Van Fossen sold his business to George Lannon; then, with a son, he sought employment as a machine operator at Bessemer. Another son, Nace, was one of the original employees of the rolling mill, but he soon tired of chipping rails and he transferred to the nail machines. In all, six brothers worked at the steel plant; for 50 years, one or more was always on the payroll.[38]

The Bessemer Works experienced its first serious labor disturbance in the nail mill. In November 1882, several feeders walked off their jobs, charging that they had been assured wages ten percent above the prevailing rate in the East, but that compensation for piece work fell far below that level. In all, 16 men were involved in the dispute. Some sought work locally or in Denver; others returned to Pennsylvania.[39] Within a month, they had been replaced and normal production had been restored.

The spike mill was adjacent to but separate from the building that housed the puddling and nail mills. Production of several sizes of railroad track spikes commenced July 10, 1882, under the direction of C. F. Drake. Equipped to yield a volume of 1,500,000 pounds, or 15,000 kegs, annually, the mill produced about one-third of that amount in the first six months of operation.[40]

Kegs for shipping nails and spikes were manufactured in a wing of the nail mill. Hoops were purchased in Kansas City, while staves and heads were cut on the premises from Colorado pine. A working force of 12 men produced up to 350 kegs daily.[41]

In order to obtain raw materials for the production of iron, in 1881 the company opened iron mines and a limestone quarry and enlarged the El Moro coking plant. All of the essential resources were found in Colorado and, with the exception of some ore, came from properties owned or leased by the corporation.

The magnetite on Grape Creek was at first inaccessible, and the cost of constructing a railroad to the deposit was prohibitive. Therefore, no attempt was made to exploit that potentially rich area until 1882, when about 900 tons were hauled by wagon to Canon City and shipped by rail to Bessemer.[42] Although a small amount was used in the puddling mill's cupolas, the ore's high titanium content precluded its reduction in the blast furnace.[43] The mine was closed in 1882 after rains had washed out the wagon road to the site.

The company augmented its iron resources in 1880 by leasing deposits on the Trinchera Estate near Placer Station in the San Luis Valley, and by purchasing the Calumet, Hecla, and Smithville claims northeast of Salida. The following year negotiations were completed for the acquisition of the Hot Springs deposits near Villa Grove.[44]

The Placer Mine, on the west slope of the Culebra Range in Costilla County, was opened in 1881 by Edward Salteel, a mining entrepreneur, for the coal and iron company. That year, the Denver and Rio Grande constructed a narrow gauge branch to a point two miles from the iron deposits, and the contractor built to the railhead through rugged terrain a tramway over which mules hauled ore-laden cars. Shipments in 1880 amounted to 1,884 gross tons and reached a peak of 7,544 tons two years later.[45] The property was abandoned in 1883 because, according to General Palmer, the cost of the ore was prohibitive. It seems more likely that that action was prompted by the inferior quality of the ore. Charles M. Rolker, a well-known mining engineer who was commissioned by the company in 1884 to evaluate all of its mineral lands, concluded that the deposit was worthless.[46]

The South Arkansas or Calumet Mine, located 19 miles northeast of Salida in Chaffee County, was a major source of ore for the Bessemer Works. Two parallel deposits were opened by a shaft, and shipments commenced in December 1881, after the Denver and Rio Grande completed an eight-mile branch to the mine.[47] Because of the remoteness of the site, the Colorado Coal and Iron Company built the town of Calumet with boarding houses, dwellings, a store, and a school to support workers and their families. Initially, wages were higher than the region's prevailing rates for miners and drill operators.[48]

The Hot Springs deposits, composed of magnetite and brown hematite, were located on the west slope of the Sangre de Cristo Mountains a thousand feet above the San Luis Valley and eight miles east of Villa Grove. The Orient Mine, opened at an altitude of 9,000 feet, was the highest iron ore producer in the United States.[49]

Production began in late 1881 when a force of 40 men opened a vein along an outcrop on the side of the mountain. Later, a slope was driven to gain access to several levels of the ore body. Large-scale shipments commenced in the following year when the Denver and Rio Grande pushed its narrow gauge line over Poncha Pass, linking Villa Grove and Mears Station near Salida. Even then, wagons were used to haul ore from the mine to the railway. It was several years before a spur was constructed to the iron deposits.[50]

The Colorado Coal and Iron Company planned to extract all raw materials for the manufacture of iron from its own lands. However, because iron ore was not immediately available, it contracted as early as November 1880 with the Breece Mining Company of Leadville for a limited supply. The purpose of that agreement was to accumulate a large stock with which to inaugurate operation of the blast furnace while proceeding with the development of mines at Placer, Calumet, and Villa Grove. Eventually, officials stated, the Bessemer Works would be substantially independent of outside sources.[51]

For many years, in fact, ores from Leadville were combined with the products of Calumet and Orient. Furthermore, the precious metal mines of Lake County were a major source of ferro-manganese ore which was used to make spiegeleisen. That material was added to the Bessemer vessel at the conclusion of the "blow" to remove excess oxygen and to regulate the carbon content of the steel. Until the company developed its own supply, it imported spiegel iron from Germany at a cost of about $200 per ton.[52]

Limestone was found near Pueblo. In 1881, the company purchased a quarter section near San Carlos Station, six miles south of the steel works along the line of the Denver and Rio Grande, on which a quarry was opened to supply fluxing material for the blast furnace and local smelters. A contractor was paid a fixed price per ton for quarrying, crushing, and loading stone on railway cars. A large quantity of clay was also discovered on the land. Kilns were erected there to manufacture brick for the construction of buildings at the steel works. For many years, that activity was carried on under the supervision of J. K. Dempsey.[53] At Bessemer, fire brick was also made for the lining of the furnace, cupolas, and converters.

In 1879, the Colorado Coal and Iron Company authorized the construction of ovens at El Moro to increase the number there to 250. At the same time, a Stutz washer was erected to remove impurities from the Engleville (El Moro) Mine's slack before it was charged in the ovens. While this reduced the ash content of the coke, which had averaged about 18 percent, the loss of fine coal in the washery increased the cost of the final product, which did not sell at a higher price. For this reason, the experiment was abandoned in 1883.[54] Almost one-half of the 108,682 tons of coke produced that year went to the steel plant.

Additional ovens were erected at Crested Butte to supply the demand for coke at Leadville and other mountain communities. In the fall of 1881, a few hundred pounds of coke were produced in open pits as an experiment to determine the suitability of coal from the newly opened Jokerville Mine for making that product. When smelter operators pronounced it satisfactory, the company commenced construction of 50 ovens. Subsequently, that number was increased to 154. As early as 1883, large shipments were made to Utah and Nevada, and one year later the market was extended to Montana.[55]

The Bessemer Works, although amply supplied with raw materials, suffered from serious handicaps. For one thing, with only a single blast furnace, the plant could not be worked at full capacity, with the result that operating costs were unusually high. In part for that reason, the company experienced difficulty in meeting competition from eastern manufacturers in a marketing region that should have been protected by high freight charges.[56]

Furthermore, the fate of the Colorado Coal and Iron Company was intimately tied to the Denver and Rio Grande Railway, the largest consumer of the Bessemer Works' steel products and the principal carrier of its raw and finished materials. Although Palmer was president of both enterprises, their relations were not good. The railway persistently found fault with the quality of the goods it purchased from the manufacturer, from whom it demanded immediate payment of transportation charges in cash while it bought what it needed on credit or with depreciated securities.[57]

For a time, the iron trade flourished, but the manufacture of steel rails was at best a marginal operation. In November 1882, Danforth warned General Palmer that a decision had to be made soon about the future operations of the rail mill. If it was shut down, as contemplated, new products had to be introduced at Bessemer to consume the daily output of the blast furnace.[58]

A depressed market for iron and steel products in 1883 posed a serious threat to the already shaky western enterprise. That year, the company elected to operate the Bessemer Works and, if necessary, to sell the output at a discount in order to dispose of surpluses. Because of the uncertain demand for pig iron, work was suspended on the second blast furnace. When "Betsy" was shut down for major repairs in December, the converter and rail mill, deprived of raw material, also closed.[59]

That year, the output of the Bessemer Works was small. The rolling mills ran at about one-half of capacity, producing rails principally for the Denver and Rio Grande and the Chicago, Burlington and Quincy. Only the nail and spike mills recorded large volumes. The Denver Rolling Mill found a ready market for its light rails and merchant iron.[60]

The prospect of losses generated widespread unrest within the board of directors, whose membership had been expanded in 1883 from five to nine members. Elected to those positions on May 28 were Palmer, Robert H. Lamborn, Danforth, Dodge, C. F. Woerishoffer, August Rutten, William H. Hays, Henry Amy, and John A. Scott, the majority being eastern financiers. Palmer was reelected to the presidency, while Lamborn and Danforth were renamed vice-presidents.[61]

The new board, reacting to the company's deteriorating financial condition, quickly assumed an active role in the day-to-day administration of the Bessemer Works in an attempt to reduce operating costs. This interference must have been resented by Palmer who, on August 8, while in the midst of a fight for control of the railway, submitted his resignation as chief executive officer of the coal and iron enterprise. No action was taken until October, when the board asked that the president reconsider. The next day, Palmer withdrew his resignation on the grounds that the corporation should not risk an interregnum in its leadership, but he implied that he would step down when a decision had been made on his successor.[62]

General Palmer's prestige was further undermined in November 1883, when the federal government won the initial round in its suit to reclaim valuable mineral fuel properties that had been acquired by the Colorado Coal and Iron Company from its predecessor, the Southern Colorado Coal and Town Company. The United States Attorney for Colorado had challenged one tract near Engleville as early as 1878 and had reinstituted a suit to invalidate the patent three years later.[63] The case remained dormant until 1881, when the Department of Justice, on request from the Department of the Interior, filed a single suit to cancel 61 patents in Las Animas and Huerfano counties, charging that approximately 10,000 acres had been taken illegally from the public domain.

In the *United States vs. The Colorado*

Coal and Iron Company,[64] the prosecution charged that officials of the District Land Office in Pueblo had conspired with A. C. Hunt and others who formed the Southern Colorado Coal and Town Company to defraud the government of a large number of tracts of land by employing the names of fictitious entrymen and witnesses to acquire rich coal deposits as agricultural lands. The plaintiff insisted that since the patents had been secured by mythical persons, no titles had in fact passed from the government. For that reason, the proceedings centered about the character of the preemptors and of the witnesses who had supported their claims.

The evidence was indirect and negative. It revealed that the entrymen could not have been real people. Fourteen witnesses testified that the several tracts in question had not been occupied, improved, or cultivated; that the preemptors had not resided on their claims as required by law; and that they were unknown to residents in the vicinity.

Attorneys for the defendant countered with four arguments. First, they insisted that the testimony had not supported the charge that the patentees were myths and that for that reason no titles had passed from the government.[65] Second, if the lands had been secured by fraud as charged, the action had been perpetrated by public officials; therefore, as an established rule of equity between two parties, the one who committed an injury must be made to suffer. Third, in the nine years since the patents were issued, the lands had passed into the control of a bona fide purchaser who had expended large sums in their development. Finally, because the government had received the full statutory price for the lands, no pecuniary fraud had been committed; the government could not retain the money and at the same time act to disaffirm and cancel the patents and to recover the lands.

The United States Circuit Court in 1883 upheld the complainant, ruling that the lands had been acquired by fraud because the preemptors were myths. Since the patentees were fictitious, no titles had been legally conveyed from the United States; there could be no transfer of titles where there were no recipients. Therefore, the Colorado Coal and Iron Company's titles were absolutely null and void.[66]

The court struck down the defendant's claim that it was a bona fide purchaser. That defense, the judge ruled, could be maintained only if the company had received legal titles. Since there had been no grantees, titles had not passed from the United States, and for that reason the doctrine of innocent purchaser could not be sustained.

An adverse decision had not been expected, yet the company had taken steps to protect against that possibility. In the summer of 1882, the fuel department had inaugurated improvements on the more valuable undeveloped coal lands. If necessary, it planned to file on the best of the mineral deposits under the laws that applied to coal lands and to allow the remainder to revert to the government.[67]

Following the decision, the company's board of directors authorized the vice president and general manager to take whatever steps were necessary to secure titles to the commercially valuable coal lands near Trinidad. Thirty thousand dollars was allocated for that purpose.[68] In this way, the corporation could retain at most only 1,500 of the 10,000 acres of the disputed lands it had acquired from the Southern Colorado Coal and Town Company.

Pending an appeal of the decision to the United States Supreme Court, the Colorado Coal and Iron Company continued to mine coal in Huerfano and Las Animas counties. A bond was posted with the Circuit Court to assure the government full indemnity for all fuel extracted from the lands if the appeal should be denied.[69]

The company faced a new crisis in January 1884, when Frederick Lovejoy, Palmer's successor as president of the Denver and Rio Grande Railway, revoked the contract under which the Bessemer Works had enjoyed special rates on the transportation of raw materials and finished products. The directors immediately protested the change in policy which had enabled the two enterprises to work "harmoniously and to mutual advantage."[70] David C. Dodge, the general manager of the railway, called the action suicidal in that it severed relations with the carrier's principal customer.[71]

For several months, the two companies waged a battle of words in the state's newspapers, each side charging the other with action detrimental to its interest. The railway insisted that rebates to the coal and iron company, guaranteed by the contract of 1880, had been excessive, in one instance amounting to almost 90 percent of the transportation charges.[72] The iron and steel manufacturer denied that it had benefitted unduly from the arrangement. To support this point, officials noted that they had paid to the Denver and Rio Grande a total of $140,450 for services in one month, an amount equal to more than one-third of the carrier's total freight business.[73]

Furthermore, the coal and iron company insisted that because of the new discriminatory rates imposed by the railway, it had been unable to market iron and steel products in Utah, Nevada, and California, while that region was readily accessible to eastern manufacturers. The Pueblo firm claimed that for shipments of nails to Salt Lake City and Ogden, it had been charged under Lovejoy's policy the "regular tariff rate" of $1.48 per keg while competitors paid only $.11 for equivalent service.[74]

The Denver and Rio Grande's motives, according to Palmer, were obvious. In seeking a so-called "equitable readjustment" of agreements, it hoped to extend its influence.[75] The New York bankers who had seized the railway's management wanted a dominant voice in the affairs of the Bessemer Works and related mining operations. If the stockholders allowed a transfer of control, the General warned, the carrier would gain absolute power over the heavy industry in that it could fix prices for fuel, rails, and other supplies, limit the company's trade, and subserve interest elsewhere at the expense of the investors.[76]

An anti-Palmer faction sought to discredit the management, charging that it had deliberately concealed "by a system of plausible reports and promises" the enterprise's financial condition. At a special meeting of investors in New York in late March 1884, the dissenters, led by Thomas E. H. Curtis requested the appointment of a committee to determine among other things whether the officers had closed mines, ovens, and mills in order to make it appear that the Denver and Rio Grande was discriminating against them, when, as the railway claimed, it was not, and if by closing the Bessemer Works, the stockholders had been injured.[77] The investigators were to make a report as to when and how the company's stock had been originally issued and what compensation had been received for it. They were also to examine the manner in which coal lands had been acquired, their cost, and the amount for which they had been capitalized, from whom they had been purchased, and "what amount of profit, if any, W. J. Palmer, W. A. Bell, William S. Jackson, A. C. Hunt, R. H. Lamborn and any others realized from such transactions." Amid confusion and without parliamentary order, the request was voted down by a show of hands. Palmer had avoided public censure, but it was apparent that his influence had been effectively undermined.

Less than two weeks prior to the corporation's annual meeting, scheduled in early April, Palmer's friends organized joint committees in Philadelphia and New York to muster support for a slate of directors. On that ticket were Edward Lewis and F. A. Dick of Philadelphia; John A. Small of York, Pennsylvania; Henry Amy, Rossiter W. Raymond, George F. Peabody, and Charles A. Moore of New York; and D. C. Dodge and A. H. Danforth. Lewis was suggested for the presidency.[78]

When the stockholders convened in Colorado Springs, the strengths of the rival factions were not immediately apparent because of delays in the validation of proxies. No business was conducted on April 7 while the tellers sorted out conflicting claims and certified voting shares personally in attendance.[79]

On the morning of April 8, the anti-Palmer forces were clearly in ascendency. The first order of business, approval of the president's annual report to the stockholders, afforded the majority an opportunity to reveal its strength. When Lyman K. Bass moved for acceptance of the document, objections were raised to what was traditionally a routine matter. By a vote of almost two to one, the report was referred to the incoming board of directors for appropriate action.[80]

The outcome of that issue was determined by Henry E. Sprague, a supporter of Lovejoy; Sprague had been elected to the directorate of the Denver and Rio Grande Railway the previous day. In person and by proxy Sprague controlled 50,708 votes. The main strength of the minority faction rested with Dr. Bell and W. S. Jackson, who together held less than 30,000 shares. The anti-Palmer forces experienced no difficulty in imposing their will on the meeting.[81]

The new board of directors for the coal and iron company was chosen by Sprague. Joining him on that body were Coleman Benedict, John E. Devlin, William A. Dick, W. B. Lawrence, August Rutten, and Thomas F. Ryan, all of whom had been hostile to Palmer's administration; members John A. Small and Danforth were representatives of the minority.[82]

After surrendering the management of the railway and heavy industry in Colorado, Palmer devoted his attention to enterprises he had initiated in Utah and Mexico. He remained president of the Denver and Rio Grande Western Railway until 1901, when George Jay Gould purchased controlling interest and merged that line with the Denver and Rio Grande Railroad.[83] By that date, Palmer had lost control of the Mexican National Railway, but he retained a large interest which he sold in 1902.[84] In the last years of his life, Palmer gave up all business activities, preferring to use his time and wealth in promoting numerous philanthropies which fostered the growth of Colorado Springs and the region.

General Palmer's successor as the president of the coal and iron company faced the difficult task of revitalizing a heavy industry that had lost a major share of the regional market on which it was dependent for prosperity. The Bessemer Works, with supporting mines and ovens, constituted an investment of nearly two million dollars.[85] Profits had been deficient and dividends nonexistent. There was little reason to anticipate a change in the corporation's fortunes unless it could reduce the costs of metallurgical products to levels that permitted effective competition with eastern rivals in the western market.

Walsenburg Area Camps

Wash house at Walsen Mine, 1915.

Power house at Walsen Mine, 1915.

The Walsen-Robinson slopes, located at Walsenburg, were the oldest and largest coal producers in Huerfano County. The Walsen mine was opened in 1878 by A. C. Hunt for the Southern Colorado Coal and Town Co. Nine years later, the Colorado Coal and Iron Co. opened on adjacent land the Robinson slope. When the underground workings came together, the two mines were worked as a unit. Production for the period from 1878 to 1931 exceeded 12 million tons.

"Colored club house" at Walsen. Black employees lived in a separate section of the community and had their own social club building in this camp, one of the few where large numbers of Negroes were employed in the mines. 1915.

An advertisement of the Colorado Supply Co. on the ball park fence at the Walsen Mine, 1915.

Barn and warehouse at Walsen, ca. 1915.

Miner's home at Walsen, ca. 1910.

Tabasco Mine, 17 miles northwest of Trinidad and one mile west of Berwind, began producing coal in 1900. A washery prepared coal for consumption in ovens which began operation at Tabasco in 1901. The beehive ovens were shut down in 1918, the year by-product coke ovens began operation at the steel works. Thereafter, Tabasco coal was sold to railroads and to domestic consumers. The mine was closed in 1930, after having yielded 4.4 million tons of coal.

Trinidad Area Camps

CF&I officials standing in front of the blacksmith shop at Tabasco, ca. 1915. From left to right are Starr J. Murphy, vice president; J.P. Thomas, superintendent of southern division mines; H. E. Weitzel, manager of the fuel department; and S. J. Pierson, vice president and general purchasing agent.

The Engle Mine, originally called El Moro, was opened in 1877 by the Southern Colorado Coal and Town Co. four miles southeast of Trinidad. For many years this was the largest mine in Colorado. In the town of Engle, or Engleville, John Tarabino owned and operated a general store until 1905, when he sold out to the Colorado Supply Co. The mine remained in operation until 1924, producing more than 8 million tons of metallurgical coal, most of which was made into coke at the El Moro ovens.

Starkville, four miles south of Trinidad, was founded as a mining and coking town by the Trinidad Coal and Coking Co., a subsidiary of the Santa Fe Railroad. That carrier eventually leased all of its mines and ovens in the Centennial State to the Colorado Fuel and Iron Co. From 1896 until closed in 1922, the Starkville Mine produced 6.5 tons of coal.

Berwind store, June, 1915.

Berwind Mine, also known at one time as Road Canon Mine, was named after Edward J. Berwind, president of the Colorado Coal and Iron Co. at the time the property was opened in 1890. The drift remained in operation until 1920, yielding nine million tons of coal, much of which was transformed into coke at ovens in nearby Tabasco.

Berwind Mine office, June, 1915.

Berwind School, Oct. 1915.

Miners' dwellings at Berwind had to compete with the railroad, tipple and supporting structures, and the schoolhouse (right) for space in the narrow canyon. Notice the neat row of outhouses on level ground scraped from the hillside behind the houses. On the distant hill is the tank that supplied the community with water.

Row of miners' dwellings at Sopris. This camp, six miles southwest of Trinidad, had been founded by the Denver Fuel Co. in 1887 to mine coal and to manufacture coke for Colorado's burgeoning smelting industry. The mine consisted of a drift and a slope. By 1928, when operations were halted, the property had yielded nearly nine million tons of coal, much of which was converted into coke to supply smelters and the steel works at Pueblo.

Employees' club house, Sopris, Oct. 30, 1915.

Sopris grade school, Oct. 30, 1915.

Coal miner's house at Sopris, ca. 1915. Rent for this dwelling was two dollars per room monthly. In an effort to improve social conditions in the mining camps, the Colorado Fuel and Iron Co.'s sociological department built many homes for miners, usually structures with four and five rooms.

6

The Colorado Coal and Iron Company, 1884-1892

Henry E. Sprague's succession to the presidency produced no long-term improvement in the coal and iron company's financial condition. In 1884, the deficit exceeded $80,000, most of which was attributed to the Bessemer Works and the failure of the Denver and Rio Grande to pay interests on its bonds. In October, a general labor disturbance in the southern field interrupted the coal and coke department's normally profitable operations.

The converter and rail mill were shut down in May and remained closed the rest of the year. That action was taken because competition from eastern manufacturers had undermined the price of steel rails in the West. They sold for as little as $26.50 per ton, when the actual cost of producing them at Bessemer was nearly twice that figure.[1]

Increased sales of iron products only partially compensated for the depressed market in rails. The nail mill operated at capacity, and a new product, cast pipe, was manufactured in the foundry. A merchant mill purchased from the Ogden Iron Works was erected next to the rail mill.[2] When it began turning out light rails and merchant iron, work was curtailed at the Denver Rolling Mill, because it was uneconomical to maintain similar plants in two cities.

The company's financial crisis was aggravated by the loss of more than $100,000 on the bonds of the Denver and Rio Grande and the Denver and Rio Grande Western Railways, many of which had been acquired in payment for supplies. Sprague criticized Palmer for having held on to the securities at a time when they were obviously declining in value, and viewed this as one more example of how the previous management had administered the company in the interest of the railways.[3]

In an attempt to halt the company's mounting deficit, in 1884 the board of directors ordered a reduction in the price of mining coal in Fremont County. That action was justified on the grounds that Canon coal, because of its cost, was not competitive with fuels mined elsewhere in the state, and that the Canon City Coal Company, a major rival, had already ordered a wage reduction at its mine.[4] The new policy precipitated a series of strikes that disrupted coal production for several months.

Prior to 1884, there had been several minor labor disturbances in Fremont County. Brief strikes occurred at Coal Creek from 1872 to 1877 in unsuccessful attempts to prevent reductions in the price for mining coal. In that time, it declined from $1.50 to $0.75 per ton of 2,400 pounds.[5] Walkouts also took place in 1879 and 1881 in support of demands for higher compensation.[6] The first of these was partially successful, because the company advanced the rate to $1.00 per ton. In 1882, workmen protested unsuccessfully a new policy of paying monthly wages with checks rather than in cash.[7]

Another walkout, in November 1883, was intended among other things to force the company to implement certain provisions of the state's first law for the regulation of coal mines. An act of the General Assembly, approved February 24 of that year, assured miners "full right of access and examination" of scales employed in weighing coal as it came from mines. The strikers insisted that this measure entitled them to hire their own check weighmen. The local superintendent conceded that point but refused to make other concessions. After five days, the men returned to their jobs.[8]

The strike of 1884 was prompted by an attempt to reduce wages. On July 12, Sprague, at the suggestion of the board of directors, instructed the general manager to lower wages in Fremont County by 10 percent as soon as sufficient coal had been stockpiled to meet the possibility of a walkout. On August 6, Sprague ordered immediate implementation of the new policy.[9] Two days later, notices were posted, at which time miners at Coal Creek and Oak Creek left their jobs, joining men from Rockvale who were resisting a similar reduction by the Canon City Coal Company.[10]

The wage issue offered miners an opportunity to air accumulated grievances. The strike became a demonstration in support of many reforms. For example, the men demanded a change in the weighing system to require payment for a ton of coal weighing 2,000 pounds. Also, they called upon the company to cease the practice of selling illuminating oil and blasting powder at high prices and to halt monthly exactions of one dollar from each employee for the hospital fund.[11] Objections were voiced against the "iron clad contract" by terms of which employees could not join a union, and the "blacklist" that made it almost impossible for a discharged worker to find another job in the district.[12]

At a mass meeting on August 18, strikers from Coal Creek, Williamsburg, and Rockvale overwhelmingly endorsed a plan for arousing sympathy for their cause. Delegates were sent to mining camps from Erie to Durango and from Crested Butte to Trinidad with requests that miners from throughout the state send representatives to a convention in Pueblo for the purpose of discussing the "condition of wages and labor."[13] That action threatened to extend the walkout to the northern field, elsewhere in the southern counties, and even into New Mexico Territory.

Miners from Erie, Louisville, and Walsenburg, in addition to the Fremont County camps, convened August 28 at the Odd Fellows' Hall in Pueblo to formulate the grievances of workers against the operators. During a two-day meeting, the "Coal Miners Protective Association of Colorado" was formed to unite coal diggers in support of certain objectives, namely the settlement of all disputes between management and labor by arbitration and conciliatory methods, as well as new laws to abolish the truck system, compulsory medical programs, and all contracts that infringed on individual rights.[14]

Until October, the coal miners' strike was confined to Fremont County. On the 20th of that month, workers at Walsenburg left their jobs. After a series of conferences with Danforth and John Cameron, the men returned to the mines on the 25th. Three days later, in response to a directive from the District Assembly of the Knights of Labor in Denver, they walked out again, and on this occasion were joined by the men at Engleville and Starkville, along with miners in Boulder, Jefferson and Weld counties.[15] Only the mountain district, including Crested Butte, was unaffected by the strike order.

At the direction of the Knights of Labor, delegates from mines convened in Denver on October 27 to seek a settlement of the strike. The next day, the convention invited the operators to join them "to mutually discuss and friendly deliberate upon a plan by which all . . . troubles can be speedily and permanently adjusted."[16] That meeting of labor and management led to the formation of a conciliation board composed of miners and operators which quickly reached an agreement for ending the walkout.[17] However, the settlement did not apply to all mines, because the state's largest coal producers — the Colorado Coal and Iron Company, Union Pacific Coal Company, and the Atchison, Topeka and Santa Fe Railroad — refused to participate on the conciliation board or to recognize its decisions.[18] Those three enterprises accounted for seven-eighths of all the fuel mined in Colorado.

The coal and iron company hoped to make a settlement without recognizing the Knights of Labor. On November 11, Danforth and Cameron met in Trinidad with two representatives from each of the corporation's mines in the southern field. The negotiators accepted the principle of collective bargaining and a modification of the medical program, and it was agreed that the miners should return to work pending arbitration of the wage issue.[19] Production of coal was resumed for two or three days before the men walked out again, this time on orders from the Knights of Labor. That organization denounced the so-called Trinidad agreement because the company refused to cooperate with the conciliation board in Denver.[20] The corporation and some miners viewed the new strike as an attempt by operators in the northern field to prolong the dispute, for in the absence of competition from the southern counties, they enjoyed high prices for lignite in the Colorado market.[21]

Strife continued in the southern field until mid-December. On December 20 miners employed by the Colorado Coal and Iron Company and the Santa Fe, without consulting the conciliation board or the Knights of Labor, accepted the Trinidad Agreement and a 10 percent reduction in wages.

Labor peace was short-lived in Huerfano and Las Animas counties. In mid-January, approximately one-third of the miners at Walsenburg and Engleville left their jobs in protest of the company's failure to institute at those mines the wage scale adopted at Coal Creek.[23] Within a week, workers had been recruited and production resumed under the protection of an armed guard.[24] Although the walkout ended on February 1, most of the strikers were not reemployed. Later that month, a train load of Negroes from Tennessee were assigned to the two mines.[25]

The District Assembly of the Knights of Labor, reacting to the employment of strikebreakers, called an "Inter-State and Territorial Convention of Coal Miners' Amalgamation." At Pueblo in early March, delegates from mines in Colorado, Wyoming, Utah, and New Mexico explored ways of counteracting the aggressive policy of the Colorado Coal and Iron Company, which, it was charged, had introduced a system of serfdom by replacing 300 experienced miners at Walsenburg and Engleville with Negroes.[26] The group formally condemned Danforth's hostility to the union and his opposition "to the principle of arbitration being applied in any practical and effective manner in the settlement of labor disputes."[27]

With that salvo, the Knights of Labor ended, for all practical purposes, the struggle to forge an organization of coal miners in southern Colorado. At least outwardly, harmony prevailed between the Colorado Coal and Iron Company and its employees in the coal camps for the next several years.[28]

The final episode in the labor dispute was acted out in the spring of 1885, after all mines had resumed operations. In May, spokesmen for some of the Negro miners at Walsenburg appealed to Governor Benjamin Eaton for assistance in returning to their homes in Tennessee. They charged that they had been promised good jobs and high wages, but on reaching Colorado had been reduced to a condition of serfdom by the company.[29]

Superintendent of mines John Cameron denied that the Colorado Coal and Iron Company had been guilty of any wrong in the employment of Negro miners. He insisted that only 27 men had been employed at Walsenburg, and that about a dozen later quit work, possibly because they had been influenced by strikers at that place. The men had left their jobs of their own accord and had been paid every cent due them.[30]

One of the dissident miners was Edmund D. Davis, who subsequently brought suit against the company. He charged that while residing in Chattanooga, he was approached by two Denver residents who offered him employment in Colorado as a teamster at a daily wage considerably more than he was making in Tennessee. He accepted, unaware that he had been recruited as a strikebreaker. On reaching Walsenburg, he was compelled to dig coal under the protection of an armed guard and was warned that the strikers would kill or injure him if he left the mine. From February 19 to April 6, he received only "a small allowance for himself and family," plus food and shelter "scarcely sufficient to sustain life."[31] On the latter date, he was permanently disabled when struck by a loaded coal car, after which, lacking means to return to his home, he moved to Denver. Davis asked for damages amounting to $7,500; he accepted a settlement of an undisclosed sum out of court.

The coal strike had little impact on the steel works. On January 23, a shortage of coal caused by the walkout at El Moro forced the puddle, merchant, and nail mills to curtail operations. Six days later, the blast furnace was banked to conserve the dwindling supply of coke.[32] In less than two weeks, after the southern mines had been reopened, those departments returned to normal working conditions.

That spring, the blast furnace was shut down for an indefinite period. The decision to halt the output of pig iron was based on the fact that a surplus had accumulated during the previous year while work was suspended in the converter and rail mill. The company may have been influenced by the resistance of furnacemen to a reduction in wages. All but half a dozen were dismissed, and the furnace was banked at the end of March.[33]

The Bessemer Works recorded another deficit in 1885. From mid-May throughout the balance of that year, most departments operated at less than full capacity and some not at all. Although the converter, blooming, and rail mills reopened in July, the output of steel rails amounted to less than 2,000 tons.[34] The pipe foundry, merchant and nail mills, affected by a depressed market, earned only moderate profits that did not offset losses elsewhere in the plant.[35]

The Bessemer Works, according to company officials, suffered from the dual handicaps of excessive operating costs and limited markets. Fuel, labor, and refractories were more expensive, and freight charges on raw materials were five times higher than in Pennsylvania. In addition, the demand for steel rails was deficient; only the Denver and Rio Grande consistently purchased that commodity in the West while the Chicago, Burlington and Quincy, Union Pacific, and the Atchison, Topeka and Santa Fe looked eastward for supplies.[36]

Charles M. Rolker, a consulting engineer who was hired in 1884 to evaluate the company's operations, concluded that the western market for iron and steel did not justify operating the Bessemer Works at capacity, which was required if the corporation hoped to run it at a profit. For that reason, he argued, the enterprise had to rely on its coal and coke trade to meet fixed charges on capital investment. This it could do if granted reasonable freight rates.[37]

Ignoring the implication of Rolker's report, company officials insisted that they could manufacture and sell iron and steel products at a profit if the Denver and Rio Grande made certain concessions on the transportation of raw materials. It was anticipated that with the impending reorganization of the carrier, it would adopt rates that permitted the resumption of rail production "on a basis commensurate with the importance of the steel works."[38]

This hope did not materialize. In November 1885, work in the steel department was again halted because of low prices for rails. That month, the company postponed plans to start up the blast furnace until such time as the Bessemer Works could be operated at a profit.[39]

Two months later, the company refused a large order for rails because a loss would be incurred from their manufacture. That decision revived the issue of freight charges. The board of directors complained that it would not be feasible to resume rail production until the Denver and Rio Grande reduced the charge per ton mile on the transportation of coal, coke, ore, and limestone from \$.02 to \$.0125.[40]

Under pressure from its largest customer, in April 1886 the Denver and Rio Grande proposed a system of rebates based on the selling price of rails. The coal and iron company countered with a demand for an "absolute concession" in the form of a fixed rebate on each ton of finished materials without reference to price, or a one cent per ton rate on raw materials. When the carrier agreed to the former, the board of directors authorized Sprague to reactivate the steel works.[41]

By the summer of 1886, an improvement in the iron and steel trade permitted the company to reopen the Bessemer Works. The blast furnace was restarted in August. Although it had been banked for 16 months, the production of pig iron was reinstituted without difficulty.[42] In early September, the converting, blooming, and rail mills commenced work on rails for the Denver and Rio Grande.[43]

Resumption of operations attracted old employees and new hands to Bessemer from all parts of the country. By mid-August, applicants for jobs were as thick as grasshoppers in Kansas.[44] Houses that had been vacant for months were quickly occupied as the demand for rental property far exceeded the supply.[45]

That summer, the steel plant was the

scene of frenzied activity. Switching engines shunted into position carloads of coke from El Moro, ore from Orient and Leadville, and limestone from San Carlos. Old rails arrived almost daily from the Union Pacific to supply the merchant mill with raw material for the production of light sections.[46] A new butt mill was constructed near the puddling furnaces to manufacture merchant steel, and the scrap mill, which had been inactive for months, was opened to produce wrought iron for the foundry trade.[47] Only the spike mill did not share in the new-found prosperity; persistent low prices forced the company to gradually curtail output until all work ended before the close of the year.[48]

The revival of the Bessemer Works was due largely to a resurgence of railroad building in Colorado. The demand for rails was large, and prices were higher than in recent years.[49] In November, after completing a run for the Denver and Rio Grande, the rail mill started work on an order for the Colorado Midland Railway's Aspen extension.[50]

The Bessemer Works recorded its peak production in 1887. The blast furnace, puddling, merchant, and nail mills were active throughout the year, while the pipe foundry, converter, and rolling mills were closed only for short periods. Rails were manufactured for the Denver and Rio Grande, Colorado Midland, and Denver Terminal Railway.[51] Even then, the plant did not run at capacity. One furnace supplied only enough iron for a single shift, and inadequate facilities for generating gas and steam prevented the merchant iron and steel mills from working at the same time.[52]

Instead of paying dividends to stockholders, the company used surpluses earned in 1886 and 1887 to enhance its competitive position in the western market. Two large expenditures were authorized: the first, to acquire and develop coal and iron mines south of Glenwood Springs with the objective of manufacturing pig iron in that region; the other, to complete at Bessemer Furnace No. 2, construction of which had been suspended in 1883 because of a shortage of funds.

The decision to extend operations into the valley of the Roaring Fork River was founded on two important considerations. Rival enterprises, the Colorado Fuel and the Grand River Coal and Coke companies, were then acquiring large tracts of coal lands that created a potential threat to the Pueblo-based firm's dominance of the regional coke trade. In addition, the corporation planned to combine coke made from coal on Thompson Creek, a tributary of the Crystal River, with iron ore from Ashcroft near Aspen in order to produce pig iron. In this way, the cost could be significantly lowered, permitting the Bessemer Works to turn out rails at a price that approximated that of eastern manufacturers.[53]

In order to open coal deposits on Thompson Creek, the Colorado Coal and Iron Company organized a subsidiary, the Aspen and Western Railway, to construct and operate a narrow gauge line from the proposed mine to Carbondale, the tentative location for coke ovens and reduction furnaces. In two years, the parent enterprise advanced more than a quarter of a million dollars for the project.[54] The roadbed was graded and bridges installed, but only a small amount of track was laid before a new era of financial stringency prompted the abandonment of the scheme. The Thompson Creek Mine shipped only about a hundred tons of coal, and no ore was mined at Ashcroft.[55]

Work was resumed on Furnace No. 2 because the one working unit had to be shut down for major repairs. In 1887, the board of directors authorized an expenditure of a quarter of a million dollars for a large furnace capable of producing pig iron at an estimated saving of four dollars per ton over the older model.[56]

Although completed, the new furnace remained idle while No. 1 continued in operation until December 1888, at which time it was shut down for a thorough overhaul. The demand that year did not justify an increase in the output of pig iron. At the same time, without the second unit in operation, the company's inability to run the plant at capacity undermined, because of high costs, its competitive position in the western market.

The steel works reported a small loss in 1888. That result was attributed to "the demoralization of railroad rates ... which enabled shippers of steel, iron and coal to invade territory that naturally should be supplied from Colorado."[57] The steel department, in operation only five months, furnished rails to the Denver and Rio Grande, the Denver, Texas and Forth Worth, and the Seattle, Lake Shore and Eastern railroads. Shipments to the latter line in Washington Territory were the first from Pueblo to the Pacific Coast.[58]

Persistent low prices for iron and steel products prompted the coal and iron company to renew demands for lower rates on the transportation of raw materials. In September 1888, the board of directors authorized Sprague to suspend production at Bessemer if high freight charges precluded operating at a profit.[59] One month later, the converting, blooming, and rails mills were closed. Subsequently, it was rumored that the Colorado enterprise had joined in a concerted effort by the country's manufacturers of steel rails to curtail output and raise prices.[60]

The recurring financial crisis caused changes in the management of the coal and iron company. On the grounds that salaries were extravagant and that some officials had engaged in speculative ventures to the detriment of the corporation's interests, the directors, in mid-November, ordered Sprague to dispense with Danforth's services. Three weeks later the general manager was allowed to resign.[61] He joined a Montana smelting company as a vice president; he later returned to Colorado as the executive officer of the Rocky Mountain Smelter and the Florence Oil and Refining companies.[62]

D. N. Jones was the next man to come under fire from the New York office. In September 1888, he rejected a proposed reduction in salary that was justified on grounds that work had been largely suspended at the Bessemer plant.[63] Shortly after Danforth left the company, the directors ordered Sprague to dismiss the superintendent of the steel works. However, he declined to do so when he learned that Jones was seriously ill. Jones died a few days later, the victim of what was diagnosed as "inflammation of the brain" caused by a series of family tragedies and the pressure of his duties for the coal and iron company.[64]

T. Haskins DuPuy, Danforth's successor, guided the business affairs of the company for only five months. His short tenure was due in part to poor health and also to the fact that Sprague stepped down from the presidency in February 1889.[65] The new executive officer, Edward J. Berwind, a prominent Pennsylvania coal operator, placed a man of his own choice in charge of the enterprise. In March 1889, John Dougherty, one-time comptroller of the Denver and Rio Grande, succeeded DuPuy.[66] After six months, Dougherty was replaced by R. G. Ford, former superintendent of the Bell's Gap Railroad, Altoona, Pennsylvania. His brief tenure ended with a fatal heart attack.[67] At that time the assistant general manager, E. M. Steck, was promoted to fill the vacancy, a position he retained until the corporation was absorbed by the Colorado Fuel and Iron Company two years later.[68]

In December 1888, the management of the steel works was entrusted to assistant superintendent Reese James while the company searched for a permanent successor to the late D. N. Jones. After six months, the post was given to J. B. Nau, a French metallurgist and steel-maker.[69] He resigned in April 1889, at which time the position was abolished and the duties assumed by the general manager.[70] That arrangement must have been unsatisfactory, because in July of the same year, Independence Grove, an engineer from Youngstown, Ohio, was appointed superintendent of the steel plant.[71] He occupied that office until the emergence of the successor company.

The mining department also experienced frequent changes in leadership. In 1889, DuPuy forced John Cameron to resign as superintendent of mines. George Haddon, long-time supervisor at the Coal Creek Mine, was promoted at that time.[72] Later in the same year, John Dougherty named George S. Ramsay to that position for a term of three years. He had been in charge of the mines of the Berwind-White Coal Company, one of the largest enterprises of its type in Pennsylvania.[73]

Frequent changes in management reflected the company's inability to operate the Bessemer Works at a profit. Steel production was suspended throughout 1889, and the blast furnaces were idle most of that year. No. 1 was relined, then torn down and rebuilt with increased capacity.[74] No. 2, christened "Mary," was started in November, stimulating employment in the mining camps as well as the steel works.[75] Daily consumption of the new unit amounted to more than 600 tons of coke, ore, and limestone. In 1889, only the merchant iron mill, machine shop, and foundry operated on regular schedules.[76]

The Bessemer Works expanded production in 1890. From March to December, the steel department turned out 30- and 65-pound rails for the Denver and Rio Grande, and 75-pound sections for the Union Pacific.[77] Two shifts, under the direction of David Lloyd and George Shearer as head rollers, processed approximately 23,000 tons.[78]

Nearly all of the old employees returned when the steel works resumed operations. Some had taken temporary jobs in Colorado; others came back from such distant points as California and Washington Territory.[79] By mid-March, 800 men were employed at Bessemer, and twice that number were on the payrolls of the company's mines, ovens, and quarries.[80]

An unanticipated shortage of pig iron interrupted operations. Furnace No. 2 was shut down in early April for a new lining. No. 1, completely rebuilt, was started a few days later, but even with an enlarged capacity, it could not supply all departments.[81] Therefore, enough pig iron was purchased in Missouri, Ohio, and Pennsylvania to assure completion of all contracts for rails.[82]

In an attempt to prevent a recurrence of an iron shortage, the board of directors authorized construction of a third blast furnace. It was also needed to supply the new cast pipe factory that was completed that year.[83] In November 1890, Riter & Conley Company, a Pittsburgh contractor, commenced construction of the new furnace on a site directly north of No. 1 along the plant's western boundary.[84]

Before approving the new project, the corporation's officials had concluded that an abundance of cheap fuel and ore justified the large expenditure. Furthermore, the additional unit would enable the Bessemer Works to operate at capacity, and lower costs would stimulate sales.

The company's hopes faded in the winter of 1890-1891. During the next year, the steel works remained "practically idle" and operations were curtailed at the mines, ovens, and quarries. The reversal was attributed to the failure of the banking house of Baring Brothers, and to consequent financial disturbances during which railroads did not purchase their usual quotas of supplies for construction and renewals.[85]

During the first eight months of 1891, Furnace No. 1 produced only foundry iron for the manufacture of cast pipe. The output of that item was increased appreciably in January when production was shifted from the old foundry to a new plant.[86] Shipments that year were almost entirely to water companies in Denver.

The lull was seized upon as an opportunity to make needed repairs and improvements throughout the Bessemer plant. Some modern machinery was installed in the converter and rolling mills to increase capacity and reduce costs. The merchant mill bar train, which had sat in the open air since its transfer from Ogden, was moved into the building that formerly housed the plate mill, an event that signaled the end of the manufacture of cut iron nails.[87] The erection of a roaster to burn off sulphur impurities in hematite from the Calumet Mine quadrupled the consumption of that ore by the blast furnace.[88]

The steel department was active from August to December of 1891, during which time rails were manufactured for the Denver and Rio Grande Western, Rio Grande Southern, and Union Pacific. When those orders had been filled, all work was temporarily halted.[89] That year, the company received from the Rail-Makers' Association, the so-called "steel trust," a payment amounting to more than $270,000, presumably for coordinating its output with that of the nation's largest producers.[90]

In the first ten months of 1892, the output of the Bessemer Works returned to normal levels. All departments were active throughout most of that period. The rail mill manufactured nearly 18,000 tons of steel sections for the Denver and Rio Grande and other western railroads. The plant recorded a profit, although a small one, for the first time in four years.[91]

As long as the steel works remained under the direction of the Colorado Coal and Iron Company, the earnings were not commensurate with the investment. According to extant records, the net investment from 1880 to 1892 equalled $2,674,000.[92] In that period, gross sales averaged $981,638 per year and net earnings were $71,569, prior to deductions for fixed charges.[93] In fact, the iron and steel departments recorded profits in only five years, while the corporation, from all endeavors, compiled surpluses in all but three years.[94] Earnings of the real estate and fuel department, especially the latter, generally more than exceeded losses incurred in the manufacture of iron and steel.[95]

The real estate department administered properties that were rented or leased by the company. These included buildings and dwellings in South Pueblo, Bessemer, and the mining camps, in addition to all lands other than coal, iron, and limestone deposits. Income averaged about $11,000 per year after deductions for maintenance, repairs, and taxes.[96]

The sale of real estate was an important source of income. Throughout the decade of the eighties, the company marketed town lots in South Pueblo, Bessemer, Canon City, Labran, Walsenburg, and El Moro, as well as agricultural and grazing lands in the Arkansas, Cucharas, and Purgatoire valleys.[97]

Large holdings of agricultural lands prompted the Colorado Coal and Iron Company to play an active role in the development of irrigation facilities. In 1885, the enterprise completed a survey for a ditch to carry water from the Arkansas River to several thousand acres in the vicinity of South Pueblo.[98] Construction plans were nearly at the point of implementation when available funds were diverted to the purchase of coal tracts in Pitkin and Garfield Counties, and to underwrite the Aspen and Western Railway.

The irrigation project was revived in 1888. That year, A. H. Danforth, William L. Graham, and James B. Orman, acting for the coal and iron company and other owners of land near Pueblo, organized the Bessemer Ditch Company for the purpose of constructing a canal to transport water from a point 18 miles west of the city to the mesa beyond South Pueblo.[99]

As the principal financial backer of the scheme, the coal and iron company received 7,400 shares and an appropriate amount of the ditch enterprise's mortgage bonds. One share represented the right to water for one acre. An additional 20,000 acres were to be irrigated at some future date by the construction of reservoirs into which surplus water was to be pumped each spring for distribution later in the growing season. Although the latter objective was not realized, the project opened a large tract for cultivation. Formerly valued at not more than five dollars per acre, the same property, after completion of the Bessemer canal, was worth as much as $100 per acre.[100]

The responsibility for marketing these and other lands was entrusted in 1889 to a collateral enterprise, the Colorado Coal and Iron Development Company. Originally that subsidiary had been assigned a different role. It was to have acquired and exploited coal and iron properties in Pitkin, Garfield, and Routt Counties.[101] That plan was modified to include nonmineral lands near Pueblo, and eventually the initial objective was abandoned.

The Colorado Coal and Iron Development Company was incorporated January 24, 1889, by T. Haskins DuPuy, J. Imbrie Miller, and E. G. Woolfolk. With an authorized capital of $2,500,000, the enterprise was to acquire and sell the parent enterprise's properties. Initially, this activity was restricted to Pueblo County where the new firm purchased 2,000 lots in South Pueblo in exchange for $1,000,000 of its capital stock and an equal amount of five percent twenty-year mortgage bonds.[102] In this way, the subsidiary relieved the older corporation of a large portion of its mortgage burden, provided negotiable securities for its business transactions, and stimulated the real estate market in and near South Pueblo. Sales which had averaged $50,000 per year from 1880 to 1888, increased significantly in the next two years.[103] However, income from that source declined in 1891 and 1892 in the midst of a mild but troublesome business depression.[104]

Coal and coke were reliable sources of income. From 1880 to 1892, the company owned or operated at one time or another 14 different mines which produced 7,500,000 tons of bituminous fuel. Earnings from that source amounted to $2,744,000, before deductions for depreciation and depletion.[105] In the same period, the two plants manufactured one and one-third million tons of coke and earned, before depreciation, $1,517,000.[106] One ton of coal yielded net earnings of $.36; a similar amount of coke yielded three times as much profit.

The coal and iron company had acquired from predecessors one mine in each of the three districts of the southern field. The Coal Creek, Walsen, and Engleville (El Moro) openings were the oldest commercial enterprises in the region. In the decade of the eighties, additional mines were established in Fremont, Huerfano, and Las Animas counties and at Crested Butte in Gunnison County. Elsewhere in the mountain district, valuable fuel deposits were secured, but plans for their commercial exploitation were not fully implemented.

The most important facilities were in the Trinidad district. The mine at Engleville was the state's largest producer of coal until 1890, and it was the company's principal source of fuel for the El Moro ovens. By that date, it was believed that the bituminous deposit was nearly exhausted. This problem was later resolved by the acquisition of adjacent lands.[107]

The federal government's suit challenging the validity of titles in the Engleville-El Moro area was settled in 1887 in favor of the coal and iron company. The United States Supreme Court reversed the decision of four years earlier and ordered dismissal of the action. The high tribunal based its ruling on three factors: the plaintiff had not shown conclusively that the preemptors were fictitious; the facts were not sufficient to overcome the presumption of innocence on the part of the officials in the district land office; and the existence of fraud in itself was not adequate grounds for cancelling patents if they had been conveyed to an innocent purchaser.[108]

The Gray Creek Mine, five miles northeast of Engleville, was leased in 1888 from the Gray Creek Coal and Coking Company as a replacement for the older workings which appeared to be nearly exhausted. Production at the new opening was halted after about a year because the coal did not coke as well as anticipated and because of the high cost of transportation between the mine and ovens.[109] By that time the fuel department decided it would be more economical to open a mine north of Trinidad. The Gray Creek lease was abandoned on a technicality that the lessor did not possess a valid title since the federal government was then suing for repossession alleging the land had been secured by fraud from the public domain.[110]

In June 1889, the coal and iron company obtained by lease and purchase a large tract of excellent coking coal in Road Canon 11 miles northwest of the El Moro ovens. The Denver Fuel Company had also hoped to gain possession of that land.[111] The task of opening what was initially called El Moro No. 2 commenced in May 1890, and coal was shipped the following October.[112] The name was later changed to Berwind Mine in honor of the man who was the company's president at the time the property was developed.

Four openings were worked in the Walsenburg district. The Walsen, obtained from the Southern Colorado Coal and Town Company, was at first the largest producer, but it was surpassed in 1891 by Walsen No. 2, better known as the Robinson Mine,[113] which had been opened four years earlier. The Cameron Mine, from which shipments began in 1883, was adjacent to the Walsen No. 1 and utilized its machinery in hoisting coal.[114] Work in the Cameron seam, lowest of the three strata in the field, was halted in 1882.[115] The Santa Clara Mine, located on leased land south of Walsenburg, was developed in 1889 and abandoned two years later after the company had driven slopes to all three seams. The product was neither coking coal nor domestic fuel and was not readily marketable.[116]

Four mines in Fremont County produced domestic fuel. The Coal Creek slope, acquired from the Central Colorado Improvement Company, remained in operation until the deposit was exhausted in 1889. Two years later, because of a large demand for fuel, the support columns were removed.[117] About one-half mile to the south was Coal Creek No. 2, known as the Caldwell Colliery until it was purchased in 1886 by the Colorado Coal and Iron Company. That opening remained in operation for 17 years.[118]

On Oak Creek in the vicinity of Williamsburg, the company controlled several hundred acres, most of which had been leased and then purchased outright from Henry M. Teller and Willard Teller. Oak Creek Mines No. 1 and No. 2 were opened in 1881. The first was closed in 1883 and the second in 1886 because the fuel was inferior to the product of other mines in the district.[119]

In the mountain counties, mines were established at Crested Butte and on Thompson Creek. The product of the former was sold as domestic and locomotive fuel and was used in the manufacture of coke. The latter deposit was developed in conjunction with the unsuccessful Aspen and Western Railway. Only a small output was recorded before the property was closed.

The Colorado Coal and Iron Company had at different times two mines at Crested Butte. The Jokerville slope was opened in September 1881, on land leased from the Durango Trust.[120] That property was abandoned in May 1885, a year after 59 men had been killed by an explosion within the workings. A new drift was developed on the corporation's land one-half mile south of the original opening. Crested Butte No. 2, because it was above the water level, experienced little danger from the lethal methane gas that plagued the first mine.[121]

For a brief time, the company seriously thought of developing anthracite deposits in the vicinity of Irwin, west of Crested Butte. In 1881, 240 acres were leased and some prospecting was done.[122] However, commercial exploitation was out of the question until a railroad reached the area. For that reason, the scheme was abandoned.

The coal and iron company systematically added to its coal lands in the southern field and in the mountain district. Nearly 9,000 acres were purchased, and an even larger amount was secured by lease.[123] That policy had two objectives. It accumulated large reserves for existing mines as well as future openings, and it discouraged competition in the coal trade by bringing key tracts under the control of a single enterprise.

Competition posed a genuine threat to the coal and iron enterprise. Its position in the southern field was challenged at an early date by railroads seeking locomotive fuel and revenues from the shipment of coal and coke throughout the region. The Trinidad Coal and Coking Company and the Canon City Coal Company firmly established the Santa Fe Railway in Las Animas and Fremont counties, respectively. The former market was opened to the Chicago, Burlington and Quincy by the Denver Fuel Company's mine and ovens at Sopris. The Chicago, Rock Island and Pacific gained access to the Walsenburg area through the operations of the Southern Colorado Coal Company.

From 1886 to 1889, the largely untapped coking coal resources of the mountain district, the key to a vast fuel trade in western Colorado and Utah, were the object of a fierce struggle between the coal and iron company and two rivals, the Grand River Coal and Coke Company and the Colorado Fuel Company, affiliated with the Colorado Midland and the Chicago, Burlington and Quincy, respectively. The Pueblo firm secured more than 2,000 acres, mainly in Spring Gulch and South Canyon near Glenwood Springs.[124] However, competitors won control of the best deposits.

The Grand River Coal and Coke Company opened mines and erected ovens that threatened the Colorado Coal and Iron Company's long-standing dominance in western Colorado and Utah. The two firms agreed in 1887 to resolve outstanding differences and to restrict competition. By terms of an agreement implemented that year, they divided claims to disputed coal lands in Garfield and Pitkin Counties; at the same time, they fixed sales quotas and established minimum prices for coal and coke where their products entered common markets.[125] Two years later, the Pueblo corporation reopened negotiations in a successful bid for a pooling of business at Leadville.[126]

The Colorado Fuel Company was the greater threat. Its aggressive policy of opening new mines and expanding markets had a detrimental impact on the Pueblo firm. In 1891, the latter reportedly lost to its rival more than a million dollars worth of fuel business, including large contracts with the Missouri Pacific and the Union Pacific.[127] This may have prompted discussions that commenced in January 1892 about the possibility of a merger.[128] Rather than perpetuate a contest that might ruin both enterprises, they agreed to consolidate their fuel operations. This was formally completed October 21, 1892, creating a giant industry in the West, the Colorado Fuel and Iron Company.

Mountain District Camps

Redstone Inn, a 40-room structure modeled after a Dutch tavern, was built by John C. Osgood for single men employed at the Redstone coke ovens.

Colorado Supply Store, Redstone, ca. 1915.

Cleveholm, a 42-room Tudor mansion half a mile south of Redstone, was erected by John C. Osgood as part of the model industrial community he hoped to establish in the Crystal River Valley. Built at a cost of about $2.5 million, the structure was completed in 1903, the year that Osgood lost control of the Colorado Fuel and Iron Co. Subsequently, he spent little time at the baronial estate. He returned there in 1925 to struggle against a fatal illness and died at Cleveholm early the following year. Courtesy of The State Historical Society of Colorado.

Coal Basin, located in Pitkin County, 12 miles west of Redstone, was served by a narrow gauge branch of the Crystal River Railroad. The mine was operational from 1900 to 1909, producing one million tons of coal, most of which was converted into coke at the Redstone ovens.

Cottage in Redstone built for workers at ovens located there to make coke from coal mined at Coal Basin.

Superintendent's residence, Spring Gulch.

Spring Gulch Mine, located at the terminus of the Jerome Park branch of the Colorado Midland Railroad, 20 miles south of Glenwood Springs. The original drift was opened by the Grand River Coal and Coke Co. in 1887. It was abandoned because of an underground fire in 1899, at which time a slope was driven to another seam. The mine was abandoned in 1916. Most of the 3.5 million tons of coal produced at Spring Gulch went to the ovens at Cardiff to be converted into coke to smelters in Colorado and Utah.

Spring Gulch school house, Sept. 1915.

The Spring Gulch-Sunlight Special, April, 1915. This mode of transportation prevailed between the two camps, seven miles apart, on the Jerome Park branch of the Colorado Midland Railroad.

Misses Keating, school teachers at Spring Gulch, Sept. 1915.

The Cardiff ovens, four miles south of Glenwood Springs, were built by the Grand River Coal and Coke Co., Jerome B. Wheeler, President, to supply smelters in Colorado's central Rockies and in Utah. The plant passed into the hands of the Colorado Fuel Co. in 1891, and to the Colorado Fuel and Iron Co. the following year. Operations were greatly reduced in 1910, and suspended altogether in 1915. Total coke production was 1.3 million tons.

Colorado Supply Co. store, Spring Gulch, April, 1915.

John Renstrom's store and residence, Spring Gulch, 1916.

Crested Butte Mine was operational from 1881 to 1952, yielding 10.2 million tons of coal. For many years, more than half of the output was converted into coke in ovens located at the mine. The remainder of the production was sold to the Denver and Rio Grande Railroad, to mining companies, and locally for domestic consumption. The snowshed over the tramway between the mine and tipple permitted operation during winter months when snowfall was usually heavy.

Miners' houses at Crested Butte. Some of these three- and four-room dwellings were moved to Gunnison after the mine closed in 1952, and were converted into motel units.

The Colorado Supply Co. store in Crested Butte, 1915.

Anthracite Mine, located five miles northwest of Crested Butte, was opened by the Anthracite Mesa Coal Mining Co. in 1882. The Colorado Fuel Co. leased the property in 1884. It was worked by that firm and its successor, the Colorado Fuel and Iron Co., until 1920. The mine, opened by a drift on the side of a mountain, was connected with the breaker by an inclined tramway 1,800 feet in length. Gunnison County was one of the few sources of anthracite outside of Pennsylvania. Courtesy of The State Historical Society of Colorado.

Pershing Mine, also known as Elk Mountain, near Crested Butte.

Anthracite breaker at Floresta, 11 miles west of Crested Butte. Originally known as Ruby Mine, this property was called the highest coal mine in the world, being at an altitude of more than 10,000 feet. Operations were limited to about six months of every year because snow frequently blocked the Denver and Rio Grande's narrow gauge line to Crested Butte. The mine remained active from 1892 to 1919, during which time it produced more than 800 thousand tons of anthracite that was marketed in Colorado and throughout the West.

Floresta school house and playground.

Miners' homes at Floresta, ca. 1915.

Heavy snowfall at Floresta halted production of anthracite during the winter months of each year. In this scene, ca. 1915, a Denver and Rio Grande narrow gauge rotary plow is clearing track in the mining camp.

The Colorado Supply Company's store, as well as other structures in Floresta, had entrances on two levels, the upper one for use when the camp experienced heavy snows during the winter months.

The Colorado Fuel Company leased in 1884 from the Anthracite Mesa Coal and Mining Company, a subsidiary of the Durango Coal and Land Company, a mine and more than 700 acres from which to ship fuel to the Denver market. The property, located on the Slate River four miles northwest of Crested Butte, was served by a spur of the Denver and Rio Grande's narrow gauge branch from Gunnison.[13]

In order to exercise control over its source of lignite, a relatively inexpensive fuel that was widely used by railroads and domestic consumers, the Colorado Fuel Company purchased the Mitchell Mine near Erie, 23 miles north of Denver. On November 15, 1887, the Mitchell Coal and Land Company, a subsidiary of the Chicago, Burlington and Quincy, conveyed the property, including 555 acres, by terms of a contract that called for payment at the rate of $.05 per ton on all coal taken from the land.[14]

The Colorado Fuel Company also sought additional sources of coal in the mountain district. In the Crystal River valley of Pitkin County, the corporation discovered what appeared to be coking-grade fuel. Tom Potter, president of the Chicago, Burlington and Quincy, and a number of Denver capitalists put up the money to purchase the land and to prospect for coal.[15]

Subsidiary enterprises were organized to facilitate exploration and exploitation of the fuel resources. On June 30, 1886, Osgood, Toll, and Jerome formed the Crystal River Toll Road Company to build and operate wagon roads throughout the coal region.[16] However, no construction was undertaken at that time.

Osgood, with encouragement from the Chicago, Burlington and Quincy, also projected a railroad from the upper portion of the Crystal River to Glenwood Springs and the Colorado-Utah border. The Colorado and Utah Railway was incorporated in July 1886 by Osgood, Kebler, Jerome, George A. Easterly, and Henry Trowbridge. Its authorized capital was $1,000,000. The line was surveyed in 1886 and 1887 from Yule Creek to Thompson Creek, a distance of 24 miles, but no roadbed was constructed.[17] By the latter date, Osgood's attention had turned to the fuel deposits of southern Colorado where railroads already existed for transporting coal and coke to markets.

With the support of Colorado capitalists, Osgood organized in 1887 the Denver Fuel Company for the purpose of purchasing and developing bituminous coal deposits near Trinidad. The directors, in addition to Osgood, were William H. James, Henry R. Wolcott, Samuel N. Wood, James V. Dexter, James B. Grant, and Dennis Sullivan. Edward O. Wolcott and Edward Eddy also invested in the enterprise.[18]

In April 1887, Sullivan, acting as agent for the Denver Fuel Company, agreed to purchase more than 2,000 acres of coal land from Elbridge B. Sopris, a Trinidad businessman, at a price of from $40 to $50 per acre. Five months later, Sullivan conveyed the property to the corporation in return for 650 preferred and 5,550 common shares of stock. By prior agreement, these securities were distributed to Osgood and the other directors.[19]

The land was actually purchased from Charles D. Chase, trustee, with Sopris acting as broker. On August 11, 1887, Chase executed a quit-claim deed to his former business partner, Maurice Lindsay, who in turn transferred it to Sullivan. For his role in the transaction, Sopris earned a sizeable commission plus a share of all profits from the property.[20]

The Denver Fuel Company probably knew at the time that it acquired the lands from Sullivan that the validity of the titles had been challenged by the federal government. In the *Trinidad Daily Advertiser* of December 7, 1886, and for some time thereafter, the General Land Office had published a notice of its intention to bring suit through the Department of Justice to cancel patents for the so-called "Sopris Lands."

These coal lands, aggregating 2,229 acres a few miles west of Trinidad, had been deeded by entrymen, who had secured them under the terms of the Preemption and Homestead Acts, to A. M. Holt. From him they passed to George M. Converse, a local gambler known as "Texas George," then to Sopris, and finally to Chase as trustee for unidentified parties.[21]

The Department of Justice, on July 27, 1887, petitioned the United States Circuit Court in Denver for an order cancelling the patents. In the bill of complaint for the *United States vs. Charles D. Chase, Trustee, and Elbridge B. Sopris,* Henry Hobson, the prosecutor, charged that Sopris, Holt, and Converse had conspired to obtain by fraud 14 tracts known to contain coal deposits. The entrymen were fictitious, as were the witnesses who supported their claims. Each of the grantees had received deeds with full knowledge of the fraud.[22] Because claimants had been myths, the plaintiff insisted the titles had not passed from the government; therefore, the court was asked to return the several tracts to the public domain.

The Denver Fuel Company, as grantee from Chase, intervened against the government. The corporation insisted that it was an innocent purchaser, having had no prior knowledge of the alleged frauds, and that for that reason its titles were valid and immune from the pending legal action.[23]

The government failed to carry its case to a satisfactory conclusion because of the recent decision of the Supreme Court in *U.S. vs. The Colorado Coal and Iron Company.* According to the United States Attorney for Colorado, that ruling made it clear that no matter what frauds had been committed, lands subsequently acquired by bona fide purchasers were not subject to repossession.[24]

While the prosecution was prepared to prove that Holt, Converse, Sopris, and even the Denver Fuel Company had known of the false entries prior to gaining possession of the lands, there was no evidence to contradict Chase's claim that he was unaware of any wrongdoing. Sopris, in Hobson's estimation, was a shrewd manipulator who was too experienced in such matters to leave his tracks uncovered. His transactions with New York parties were "sufficiently complete to cover all frauds under the doctrine of innocent purchase."[25] It appeared that the men responsible for the frauds had deliberately placed the property in the hands of Chase in order to deny to the government the right of redress through the courts. At the request of the plaintiff, the suit was dismissed November 8, 1888.

The Denver Fuel Company opened a mine at Sopris. Operations commenced in November 1887, and the first coal was shipped the following January. At maximum capacity, one slope produced about 1,400 tons daily, most of which was sent to market over the Denver and Rio Grande and the Denver, Texas and Fort Worth roads. In September 1888, on completion of 100 beehive ovens, all slack coal and screenings were converted into coke for sale to the smelting industry.[26]

The mine was supervised by Kebler, who moved to Colorado in 1887 as general manager of the fuel company. The work of driving the slope, constructing the tipple and ovens, and the installation of machinery was directed by Raphael M. Hosea, chief engineer, who had occupied a similar post with the Whitebreast Coal and Mining Company.

The Sopris Mine remained in the possession of the Denver Fuel Company until April 1, 1889. During that period, the output was 224,689 tons, only about one-tenth of which was diverted to the ovens to manufacture 15,096 tons of coke. The net return on coal was about $.60 per ton and on coke, $1.23. After deductions for interest, insurance, taxes, and sinking funds, the corporation's earnings amounted to $94,829.88, from which dividends equal to 6 percent of preferred and common stock were paid to investors. The undistributed surplus was almost $27,000.[27]

In the summer of 1888, the people who controlled the Denver Fuel Company moved into the rich coal lands of Pitkin and Garfield counties. On August 24, Osgood, Kebler, H. R. Wolcott, and Beaman organized the Elk Mountain Fuel Company with an authorized capital of $2,000,000.[28] The directors for the first year were the incorporators. Osgood was elected president, Wolcott vice-president, and S. N. Wood treasurer.

The board immediately ordered payment of $300,000 of the capital stock to purchase properties. Osgood, for himself and others, offered to sell 4,300 acres in Coal and Pinon Basins, reportedly the best coal deposits in the mountain district, for $50 plus three shares of stock per acre.[29] For some reason, the former lands were not conveyed at that time. The company secured 1,560 acres in the latter area and obtained other tracts from local residents.

The Colorado Fuel Company was reorganized on April 1, 1888, as the first step toward the consolidation of Osgood's coal operations in a single enterprise. On May 11, the directors amended the corporation's charter by increasing the authorized capital from $500,000 to $5,000,000.[30] The additional stock was required to underwrite plans for expansion in the mountain district and the southern field.

Colorado capitalists controlled the expanded company. The directors were Osgood, Wood, Sullivan, H. R. Wolcott, and Toll. Osgood was president, Sullivan vice-president, Beaman secretary, and Wood treasurer.[31] Kebler remained the general manager, and Cass was named general sales agent.

On May 16, the directors authorized negotiations for the purchase of the Denver Fuel Company. That transaction was approved in February and formally executed on April 1, 1889, when, at a cost of $1,091,000 in stock, the Colorado Fuel Company acquired the Sopris Mine and all its properties, as well as other lands in Las Animas County.[32]

On May 1, 1889, the Colorado Fuel Company also acquired the assets of the Elk Mountain Fuel Company, including 2,640 acres of coal lands and other property in Garfield County. That transaction was completed through an exchange of stock worth $434,700. Seven shares were given for two of the smaller enterprise, whose outstanding securities were cancelled.[33]

As a result of reorganization and consolidation, the Colorado Fuel Company gained direct control of 5,622 acres of land, more than one-half of which was in Las Animas County. The balance, except for the small tract around the Mitchell Mine, was located south of Glenwood Springs. In the next four years, the corporation enlarged its real estate holdings to nearly 34,000 acres.[34]

On May 1, 1889, the fuel company executed a trust deed on its property to the Atlantic Trust Company as security for $1,200,000 in six percent general mortgage bonds. The proceeds of that issue were allocated to the acquisition of coal lands and the development of new mines.[35] Purchases were principally in the southern field, the Grand River Valley, and Wyoming. In Huerfano and Las Animas counties, the enterprise absorbed two important rivals.

The Southern Land Company had been incorporated in 1887 by Whitney Newton, a Pueblo lumber dealer, and by James A. Joy and George J. Dunbaugh, wholesale grocery merchants who also operated retail stores in Pueblo and several small communities throughout southern Colorado. Although Newton, Joy, and Dunbaugh were directors in the corporation, a majority of the stock was held by Appleton H. Danforth, William G. Brown, and John Cameron, all of whom were officers of the Colorado Coal and Iron Company. The authorized capital of nearly $150,000 was expended for coal land in four tracts: Lester, Rouse, Pictou, and Berwind. The first three were in Huerfano County and the last in Las Animas County.[36]

Instead of opening mines, the company adopted a policy of selling or leasing its lands to other enterprises. In June 1888, it transferred more than 2,000 acres to the Southern Colorado Coal Company for a consideration of $70,607. Later the same year, the balance of the property was leased to the Colorado Coal and Iron Company.[37]

In the summer of 1889, the Colorado Fuel Company quietly purchased controlling interest in the Southern Land Company. In September, Danforth, Brown, and Cameron sold five-eighths of the outstanding stock for $74,000. Early the following month, Whitney, Joy, and Dunbaugh transferred their interests for $44,000. On November 28, 1889, Kebler, Toll, H. R. Wolcott, and Beaman, representing the stockholders, authorized the transfer of all assets to Osgood's enterprise in return for $15,000 in cash and 1,496 shares of capital stock.[38] In this way, the latter corporation secured additional valuable coal lands in the southern field.

The Southern Colorado Coal Company had been incorporated in 1888 by James K. Robinson, Charles B. Patterson, and Lafe Pence with an authorized capital of half a million dollars for the purpose of operating mines in Huerfano and Las Animas counties. Robinson resigned his position as assistant superintendent of mines for the Colorado Coal and Iron Company to become vice-president and general manager of the new venture. The principal stockholders, in addition to Robinson, were George Coppell, an English financier with large investments in the Denver and Rio Grande, and R. R. Cables, president of the Chicago, Rock Island and Pacific.[39] The last two jointly held three-quarters of the stock.

Northwest of Walsenburg the company opened Sulphur Springs No. 1 and No. 2, also known as Lennox and Maitland, respectively, and also the Pictou Mine. Shipments began in late 1888, after which the three slopes were consolidated as a single operation called Pictou. The Rock Island road, using track leased from the Denver and Rio Grande, secured the entire output for its own consumption and for resale in Kansas.[40]

The Colorado Fuel Company inaugurated negotiations in August 1889 for the purchase of the Southern Colorado Coal Company's properties. The holdings of Coppell, Cables, and Robinson were acquired in separate transactions, and the corporation's assets were formally absorbed on February 1, 1890.[41] In this way, Osgood and his associates secured an additional 3,358 acres, including the important Santa Clara and Pictou tracts near Walsenburg and the Road Canon and Apishapa lands in Las Animas County. The purchase price was $150,000 in par value stock.[42]

The final acquisition in the southern field was the Big Four Tract near Walsenburg. That property, comprising 1,120 acres, had been purchased by the Huerfano Land Association, members of which included Osgood, Kebler, Sullivan, H. R. Wolcott, Wood, and Richard C. Hills. On October 1, 1892, the fuel company secured the property in a cash transaction for about $25,000. Each share in the pool was worth $3,670 in the final settlement.[43]

Prior to 1892, the Colorado Fuel Company purchased an additional 8,820 acres in the mountain district. The most important of nine distinct parcels were the Coalridge tract in the Grand River Valley southeast of Newcastle; Coalbasin, six miles west of the future town of Redstone in the Crystal River Valley; and the Floresta lands near Kebler Pass, west of Crested Butte.

The Coalridge property, comprising 1,853 acres, had been taken over from the Elk Mountain Fuel Company at the time of its consolidation. In the next two years, an additional 291 acres were purchased in the area. Although some prospecting had been done as early as 1887, most of the coal deposits, as revealed by attempts at commercial exploitation, were of little value.[44]

South of Glenwood Springs, the Coalbasin lands encompassed slightly more than 3,000 acres. All but a quarter section of that amount was purchased from Osgood, as agent for himself and others. This property was to have been conveyed to the Elk Mountain Fuel Company, but the transaction was not completed prior to consolidation. The Colorado Fuel Company obtained most of the property in exchange for 6,153 shares of stock.[45] The Denver Fuel Company surrendered in 1890 a small parcel of land in the same area.

The Floresta lands contained valuable deposits of anthracite. Located in the central portion of Gunnison County, the property had been claimed from the public domain in 1879 by the Osage Coal and Mining Company. Prevented from commercially developing the tract because of its remoteness from markets, that enterprise sold all assets to a group of speculators that included Mahlon D. Thatcher, prominent Pueblo banker. They opened a small mine to supply residents of Irwin until the mining camp passed from existence.[46]

Osgood and his associates gained control of the anthracite deposit near Floresta through the Denver Fuel Company. On February 15, 1890, the board of directors authorized Charles H. Toll to complete negotiations for the tract with M. D. Thatcher. A purchase contract had been signed two months earlier; all that remained to be determined was the time and method of payment. In early September, the deed was executed in the name of Julian A. Kebler, who transferred it to the corporation.[47] The property formally passed into the possession of the Colorado Fuel Company two years later.

The Denver Fuel Company had been revived on February 15, 1890, to assist the parent enterprise in securing coal lands in the Rocky Mountain region. The corporation's capital was fixed at $600,000, one-third of which was subscribed by the stockholders of the Colorado Fuel Company.[48] All proceeds were devoted to the purchase of real estate, altogether about 9,000 acres in addition to the Floresta property.

The largest of the acquisitions was in Carbon County, Wyoming. On November 1, 1890, S. W. Downey, trustee for a group of landowners, transferred 8,203 acres to Kebler for a consideration of $50,000.[49] The Denver Fuel Company purchased the title at cost plus interest, with a bonus of $10,000 in stock to Richard C. Hills for his services in making geological explorations of the area.[50] The corporation paid a total of $64,000 for the property and invested an additional $27,000 in surveys and prospecting prior to its conveyance in 1892 to the Colorado Fuel Company.[51]

By purchasing the majority interest in the Yampa River Coal Company, the Denver Fuel Company attempted unsuccessfully to open mining operations in Routt County. Organized in September 1890 by J. K. Choate, S. N. Wood, Paul Morton, and D. C. Beaman, with an authorized capital of $300,000, the coal company proposed to gain control of reportedly rich anthracite deposits in the valley of the Yampa River.[52]

Earlier that year, Choate, Wood, and Osgood had joined with C. D. Simpson and other local investors to underwrite the cost of exploring 2,000 acres said to be underlaid with coal. The association agreed to furnish money for prospecting and, if commercially valuable coal were discovered, to contribute a substantial amount to the capital of the Yampa River Coal Company.[53] Choate, Wood, and Osgood subsequently conveyed their interests to the Denver Fuel Company.[54] The venture was not pursued, perhaps because the area was not accessible by railroad. In any event, Osgood and his friends were preoccupied with other enterprises.

Prior to October 1892, the Colorado Fuel Company had in operation at one time or another a total of six coal mines. Two had been acquired from the predecessor of the same name at the time of reorganization. Anthracite, or Mine No. 1, situated at an altitude of nearly 10,000 feet above sea level in Gunnison County, was a single drift with a maximum output of 400 tons a day. Initially, the product was hauled from the workings by mules (later by a continuous rope system) and dumped into a chute that carried it 800 feet to the floor of the Slate River to a "breaker" where it was processed for shipment. Because of heavy snowfalls that interrupted railroad traffic, the mine worked only sporadically during the winter months of each year.[55]

The Mitchell Mine, in Weld County, was opened by a shaft that exposed a vein of lignite from five to six feet in thickness. The output was about 400 tons daily. In the spring of 1891, the underground workings flooded, and the property was abandoned.[56]

Mine No. 3 was located at Rouse, approximately seven miles south of Walsenburg. The task of driving three slopes commenced in April 1888, and coal was shipped in September.[57] Extensive underground workings were designed to produce 2,000 tons daily for shipment over the Denver, Texas and Fort Worth Railway. A new town established on the property contained a boarding house, a large number of dwellings, and a general store, of which all but the family units were managed by a subsidiary enterprise, the Colorado Supply Company.[58]

Mine No. 4, six miles southwest of Trinidad at Sopris, consisted of three drifts, one each to the east, south, and west along a seam of bituminous coal that cropped out at that point. The product was ideal for the generation of steam and gas, as well as for smithing and coking. Loaded cars were hauled from the openings to a tipple, a distance of 1,000 feet, by a tail rope system. After screening, the marketable fuel was loaded on railway cars provided by a spur of the Denver, Texas and Fort Worth line, and the remainder was transported to nearby beehive ovens for the manufacture of coke.

By 1890, the Sopris Mine was the largest producer in the state, surpassing that year the output of the Colorado Coal and Iron Company's mine at Engleville. The daily capacity of the Sopris openings was increased to 1,750 tons, but they yielded as much as 2,200 tons in a ten-hour day.[59]

One hundred beehive ovens at Sopris had a capacity of about 30,000 tons of coke per year. In 1892, on completion of an additional 122 units, the output was more than doubled.[60] The product was of superior quality, low in percentage of ash and with only small amounts of sulphur and phosphorous. Its dense cellular structure made it an ideal fuel for reducing iron and ores of precious metals.[61]

For a time, the Colorado Fuel Company seriously considered opening a mine in Road Canon north of Trinidad. Plans evolved to the point where, on November 30, 1888, Sullivan, H. R. Wolcott, Toll, Beaman, and Wood organized the Las Animas and Western Railway Company with an authorized capital of $100,000 to construct a line from Barnes Station on the mainline of the Denver and Rio Grande to and through Road Canon, Stock Canon, and Canon del Agua.[62] Before surveys could be inaugurated, the scheme was abandoned, probably because the Colorado Coal and Iron Company secured by purchase and lease most of the exploitable fuel resources in Road Canon for what became its El Moro No. 2, or Berwind Mine.

Coalridge, or Mine No. 5, was located southeast of Newcastle along the line of the Colorado Midland Railway on the so-called Pinon Basin lands which had been acquired from the Elk Mountain Fuel Company. An opening driven to the upper of two strata in 1889 revealed that much of the coal had been oxidized and that mining operations would be uneconomical because of a series of geological faults that ran throughout the area.[63] These features had not been disclosed by drill samples when the property was prospected. A slope driven to the lower level revealed that it was equally unsatisfactory. All production was terminated in 1892, after which the buildings and machinery were dismantled and transferred to other mining properties.[64]

Located four miles northwest of Walsenburg on the Loma Branch of the Denver and Rio Grande, the Pictou, or No. 6 Mine, consisted of three openings that had been acquired from the Southern Colorado Coal Company. The community's namesake in Nova Scotia had been the home of Tommy Lawther, who had supervised the earliest mining operations in the area.[65] The property had a daily capacity of about 900 tons of noncoking, steam, and domestic fuels from the Walsen, Cameron, and Robinson seams; most of this was sold to the Union Pacific Railway.[66]

In the summer of 1892, the Colorado Fuel Company acquired four mines in the mountain district from the Grand River Coal and Coke Company. That enterprise had its origin in the efforts of Jerome B. Wheeler and James J. Hagerman to exploit the coal resources of Garfield and Pitkin Counties. A New York businessman and owner of mining interests at Aspen, Wheeler had secured titles to a large number of mineral fuel claims, perhaps in anticipation of manufacturing coke for the region's smelting industry. Hagerman, a wealthy Milwaukee industrialist who had retired to Colorado Springs, recognized the value of that endeavor not only for its own sake but as a source of traffic for the Colorado Midland Railway, of which he was the principal promoter.

That broad gauge line ran from Colorado Springs to Leadville, with extensions projected to Aspen, Glenwood Springs, Grand Junction, and Salt Lake City. It was to traverse the valleys of the Roaring Fork and the Grand River, providing ready access, when completed, to rich coal resources. In order to encourage additional construction, Wheeler agreed to surrender 3,000 acres of fuel lands to a company which was to be formed by Hagerman and his associates as an ancillary enterprise of the railroad.[67]

The Grand River Coal and Coke Company was incorporated in the State of New York on May 12, 1886, by Samuel S. Sands, Rufus K. Terry, and Walter B. Devereaux. The enterprise, with an authorized capital of $1,000,000, proposed to acquire and operate coal and iron mines and coke ovens in Garfield and Pitkin counties and elsewhere in Colorado.[68] One month later, Wheeler conveyed his mineral claims in exchange for all of the venture's stock, and he and Hagerman advanced $25,000 as the initial installment of working capital.[69] Sands served as president; Devereaux, as general manager, was assigned the task of developing the properties. Six months later, Wheeler was elevated to the chief executive's post.[70]

The formation of a corporation to commercially exploit the coal resources south of Glenwood Springs produced a series of clashes between rival enterprises. The Colorado Coal and Iron Company, for example, dispatched armed men to guard its claims in that region.[71] After several skirmishes, the dispute was resolved when the Pueblo firm agreed to abandon the area north of Thompson Creek. A short time later, the Colorado Fuel Company relinquished its rights on Rock Creek in favor of concentrating on the exploration of deposits in Coal Basin.[72]

The Grand River company opened the Marion, Sunlight, and Spring Gulch mines along the Jerome Park Branch of the Colorado Midland. These openings had an aggregate capacity of about 1,000 tons daily and produced locomotive, domestic, and coking coals, respectively.[73]

Another mine was opened at Newcastle, 12 miles west of Glenwood Springs along the line of the Colorado Midland. A single slope, and later a shaft, exposed five seams that measured altogether more than 60 feet in thickness. Shipments of the excellent steam and domestic coal commenced in December 1888. Four years later, the railroad, at its own request, took almost all of its locomotive fuel from this property. Marion, because its product was excessively friable, was closed in 1892.[74]

The output of the four mines was sold mainly to railroads. In 1891, the Colorado Midland, Denver and Rio Grande, the Rio Grande Junction, Manitou and Pike's Peak, and the Chicago, Rock Island and Pacific purchased more than one-half of all the coal produced, with the first-named line accounting for nine-tenths of that amount.[75] The principal domestic markets were Leadville, Aspen, and Colorado Springs, in that order of importance, with the first taking about 16 percent of output. Only small tonnages were sold outside the state by the Kansas and Colorado Company, which held exclusive rights to market the product at points served by the Burlington and Missouri Railroad and the Union Pacific Railway.[76]

The coal business was of secondary interest to the Grand River Coal and Coke Company. More important was the trade in coke, for which there was a large and growing demand in western Colorado and in adjoining states and territories. Furthermore, there was strong competition for the domestic fuel market at Denver and Pueblo, which, because of location, could be supplied more economically from the state's other mining districts. Garfield County coke, by contrast, could be shipped to Leadville more cheaply than the output of ovens at Crested Butte or El Moro and could enter Denver on almost equal terms with the product of Colorado Coal and Iron Company or the Colorado Fuel Company.

A series of experiments were inaugurated in 1887 in the manufacture of coke from coal mined in Jerome Park. The initial test at the Marion Mine was unsatisfactory. When the product of the Spring Gulch Mine proved cokeable, ovens were erected at Cardiff, four miles south of Glenwood Springs. A battery of fifty was constructed in 1888; in three years, the number increased fivefold.[77]

The company adopted the Belgium or stack oven in an attempt to manufacture coke that surpassed in quality the Crested Butte product. The rectangular structures differed sharply in appearance from the domed models generally used elsewhere in Colorado. Experience indicated that the units at Cardiff were too small for economical operation and that the output, while good, was definitely inferior because of its fragile character compared to that of the rival enterprise in Gunnison County.[78]

General manager Devereaux was strongly criticized for having adopted the Belgium design. In an attempt to improve operations at Cardiff, he hired a coke-making expert from Pittsburgh to take over management of the ovens. Even then, the output remained small and the quality uneven. In 1891, additional units were erected, using the beehive design, but they were small and of faulty construction.[79]

Cardiff coke was consumed by smelters of precious metal ores which were located over a wide area. In 1891, the company produced in excess of 51,000 tons, more than one-third of which was shipped to Leadville, and only slightly smaller amounts were sent to Denver and to Utah.[80] The magnitude of this business had the effect of discouraging the construction of additional ovens at Crested Butte.[81]

The success of the Cardiff ovens posed a problem for the Colorado Fuel Company which planned to commercially develop the coking fuel of Coal Basin, situated south of Jerome Park in Pitkin County. After some hesitancy, Osgood and his associates decided that they should purchase the rival concern. In this way, they not only eliminated potentially troublesome competition, but they also acquired operating mines and ovens with established markets.[82]

In July, the Colorado Fuel Company took over the rival enterprise. After Wheeler, Devereaux, and their close associates agreed to sell a majority of the stock, the other shareholders accepted the same terms.[83] Each received for his holdings one-tenth of par value in cash, one-fourth in preferred and one-tenth in common stock. Osgood succeeded Wheeler as president, and Kebler, as vice-president and general manager, assumed the direction of the coal and coke company's business affairs.[84]

That acquisition established the Colorado Fuel Company in the position of first rank among the fuel enterprises in the West. In four years, the Denver-based firm had produced two and one-half million tons of coal from its own mines while purchasing about one-sixth that amount from other producers for resale. It had manufactured 147,000 tons of coke, purchasing at the same time one-tenth that amount for its customers. With additional mines and ovens in the mountain district, the firm clearly overshadowed all competitors.[85]

In contrast to the Colorado Coal and Iron Company, Osgood's corporation recorded large profits and paid substantial dividends to investors. Prior to October 31, 1892, net earnings exceeded $867,000, of which $751,000 were distributed, leaving at the time of the consolidation a surplus of $116,000.[86] This capacity for making money was not overlooked by rival concerns.

The two giants of the fuel industry agreed in January 1892 to discuss the possibility of consolidation, ostensibly to enable a successor enterprise to capture a larger share of the fuel market in Kansas and Nebraska. In fact, the Pueblo firm's position in the coal and coke trade had deteriorated to the point where it had little choice but to seek a union with its principal rival.

In June 1892, representatives of the two enterprises agreed to combine their properties and business operations, including the steel works at Bessemer, in a single corporation. That understanding was almost immediately rejected by Edward J. Berwind, who had been in Europe during the negotiations. However, after the Pueblo firm's townsite and agricultural lands were transferred to a separate concern that was not part of the consolidation, a plan was approved. On October 21, the Colorado Fuel and Iron Company succeeded to the functions of the Colorado Coal and Iron Company and the Colorado Fuel Company, ending a decade of rivalry in the region's fuel trade.

8

The Resurgence of the Steel Works, 1892-1903

During the first half of 1892, the Colorado Coal and Iron Company and the Colorado Fuel Company each employed special investigators to examine all aspects of the other firm's operations and properties as the initial step toward possible consolidation. Reports submitted to Osgood and his subordinates suggested that the rival concern was abundantly endowed with coal and iron lands, mines, ovens, and supporting facilities and that at a modest cost, all departments could be put in condition to operate at a profit. The financial shortcomings of the Pueblo enterprise were attributed to inept management and inefficient utilization of available capital, labor, and mineral resources.[1]

Soon after the merger talks commenced, it was reported that the steel plant at Bessemer was not to be included in the consolidation. This prompted a number of western railroad men to seriously consider the possibility of taking over the industry. According to one rumor that circulated in Denver, General Palmer and D. C. Dodge organized a syndicate to acquire the property because they were convinced that even without direct ownership of coal mines and coke ovens, the manufacture of iron and steel could be carried on profitably in the West.[2]

Initially, Osgood had planned to combine only the coal and coke operations of the rival enterprises. However, he changed his mind about the steel works after reading the reports of experts who examined that property as well as other assets of the coal and iron company. F. A. Delano, a metallurgical engineer employed by the Chicago, Burlington and Quincy Railroad; Theodore G. Wolf, general superintendent of the Lackawanna Iron and Steel Company; and Eliot A. Kebler, contract agent for the Addyston Pipe and Steel Company, insisted that the manufacture of iron and steel was not only an important adjunct of the fuel business but potentially the company's most profitable endeavor. Delano was convinced that an investment of only $100,000 was required to make the Bessemer Works "produce on an economical and paying basis"[3]

After prolonged negotiations, officials of the two companies began on July 12 a six-day inspection of the principal properties. A special train took them to Bessemer, Walsenburg, Trinidad, Canon City, Salida, and Glenwood Springs. The party included Channing F. Meek, the new president of the coal and iron company; directors Thomas E. H. Curtis, Adolf Ladenburg, and John Scott; and heads of operating departments. The rival enterprise was represented by Osgood, J. A. Kebler, Cass, Beaman, Jerome, Toll, H. R. Wolcott, and several subordinate officials.[4]

On returning to Denver, the key members of the group agreed on a tentative plan of consolidation. On the night of June 17, Ladenburg, a New York broker and principal negotiator for the coal and iron company, acted as host at a dinner in the Metropole Hotel to celebrate the forthcoming merger.[5] All properties and business operations were to be combined, and the Pueblo-based corporation was to acquire 57.25 percent of the capital stock of the successor firm.[6] Presumably, Osgood and his friends, because of their holdings in both enterprises, were to exercise a dominant voice in the new management.

Subsequently, both parties expressed dissatisfaction with the plan. Osgood wanted to renegotiate the agreement without the coal and iron company's large real estate holdings, apparently because he feared they would be a burden to the successor enterprise. At the same time, large stockholders in the Pueblo corporation complained that they had not been treated fairly by the consolidation scheme. Therefore, a new round of discussions was inaugurated.

Pending a final settlement, both sides agreed that the coal and iron company should be turned over to the management of the rival concern. The rationale for this move was that the former's subordinate officials were inefficient, and that they had been "unwilling or unable" to furnish information as to the affairs and properties of the enterprise to Meek or to officers of the Colorado Fuel Company.[7] Furthermore, if the merger failed to materialize, the Pueblo firm, by temporarily surrendering the administration, would be in a position to make sweeping changes in personnel. It seems more likely that both sides recognized that the union had to occur; all that remained to be decided were the terms.

On August 1, Osgood and his associates assumed control of the coal and iron company's operations. J. A. Kebler succeeded E. M. Steck as general manager, and with few exceptions the Colorado Fuel Company's superintendents assumed the equivalent roles in the other enterprise. Within less than a week, the general offices of the two firms were consolidated in Denver.[8]

At that point, Edward J. Berwind, chairman of the board of the coal and iron company, entered negotiations for a final settlement. According to one report, he was unhappy with the agreement and had hastily ended a European vacation to return to the United States in order to prevent Meek from selling out his and other stockholders' interests to Osgood. It was said that Berwind believed some of the Pueblo firm's assets had been grossly undervalued in the merger plan.[9]

Assisted by Meek, Ernest Thalmann, and H. K. McHarg, Berwind renegotiated the agreement with Osgood and Henry Wolcott. During the discussions, Berwind was "outgeneraled" by Osgood at every turn; the plan that was finally adopted clearly favored the Colorado Fuel Company, enabling it to complete the merger on its own terms.[10]

The two enterprises were to form a single corporation, the Colorado Fuel and Iron Company. Included in the consolidation were the Denver Fuel Company, the Grand River Coal and Coke Company, and the Huerfano Land Association, all of whose capital stock was owned by Osgood's corporation.[11] That firm and subsidiaries had engaged in the business of extracting coal and manufacturing coke, and their properties consisted of mines, ovens, and related facilities, the fair value of which could be readily determined.

The coal and iron enterprise, however, was more complex. Its assets consisted of the Bessemer Works, coal mines and ovens, iron mines, and other real estate. The production of iron and steel was dependent on the functions of the fuel department and ore-mining operations; but unrelated to those activities were agricultural and town lands, the value of which could not be accurately established. For that reason, the latter properties were excluded from the merger plan.[12]

In order that the real estate holdings should benefit the stockholders of the coal and iron company, it was agreed that all assets unrelated to operating departments were to be conveyed prior to consolidation to the Colorado Coal and Iron Development Company. These included 75,000 acres comprising "agricultural, grazing, fruit and oil lands, and suburban property and town lots in Pueblo, Canon City, Florence, Bessemer and El Moro"; all shares and bonds of the Bessemer Ditch Company; all bills receivable for deferred payments on lands; and all bonds of the development enterprise.[13]

The net value of the property was $4,970,000, in exchange for which the Colorado Coal and Iron Development Company was to issue an equivalent amount of stock. Those securities, together with others previously acquired, were to be distributed to the coal and iron company's shareholders in proportion to their investments.[14]

The remaining properties entered into the consolidation at actual values, as nearly as could be determined. The Bessemer Works was computed at replacement cost, less a reasonable amount for depreciation. Coal mines, iron mines, ovens, quarries, and mineral lands were figured at cost, minus depreciation and depletion, if appropriate. Materials, supplies, and stocks of manufactured goods were converted to current market-price cash equivalents. On this basis, the assets of the Colorado Coal and Iron Company were set at $7,101,000, and those of the Colorado Fuel Company, including the subsidiaries, at $8,393,000.[15]

The successor corporation was to assume all of the outstanding bonded indebtedness of the consolidating enterprises. This included $3,010,000 for the Colorado Coal and Iron Company, and $1,143,000 for the Colorado Fuel Company. Since the latter's preferred shares were treated as mortgage obligations, the liabilities of the two firms were practically the same.[16]

The Colorado Fuel and Iron Company was to be incorporated under the laws of Colorado with an authorized capital of $13,000,000. Of that amount, $2,000,000 was to be in the form of eight percent preferred certificates and exchanged share-for-share for the preferred stock of the Colorado Fuel Company. In addition, owners of the common stock in that enterprise and the coal and iron company were to receive $5,250,000 and $4,000,000 in common shares, respectively, leaving unissued a balance of $1,750,000.[17]

Furthermore, the new enterprise was to issue $6,000,000 in mortgage bonds. More than two-thirds of that amount was to be surrendered to the New York trustee for the purpose of retiring the outstanding bonds of the consolidating companies. The remainder, or $1,756,000, was to be placed in the treasury to underwrite expansion of business operations in the future.[18]

On October 21, the day following official approval of the consolidation by stockholders of the two enterprises, the successor company was formally organized. Osgood, Henry Wolcott, Sullivan, W. H. James, J. A. Kebler, Jerome, and Toll joined Meek, W. L. Graham, C. K. McHarg, Thomas E. H. Curtis, and E. F. Arthur in electing Cass to the thirteenth position on the board of directors, after which that body adopted the articles of incorporation for the Colorado Fuel and Iron Company.[19] That document was filed the same day with the Secretary of State of Colorado.

The election of officers revealed which one of the predecessors was to be ascendant in the new enterprise. Osgood was chosen president; Wolcott, Paul Morton, and Kebler vice-presidents; C. M. Schenck secretary; and Cass, who was quickly succeeded by Jerome, treasurer.[20] All of these men had been prominently associated with the Colorado Fuel Company.

Most of the company's varied business activities were entrusted to Osgood's long-time friends. General manager Kebler was assisted by Cass as general sales agent, Beaman as general attorney, Schenck as general auditor, and M. S. Donnelly as purchasing agent. Heading the operating divisions were W. P. Thompson, superintendent of the fuel department; T. B. Butler, iron mines; and Theodore W. Robinson, steel works. Dr. Richard W. Corwin, chief surgeon, had charge of all medical programs in Pueblo and the mining camps.[21]

From the predecessors, the new corporation acquired abundant resources with which to engage in mining and manufacturing activities. Fuel lands in the amount of 69,000 acres contained an estimated 400,000,000 tons of steam, domestic, coking, gas, smithing, or anthracite coals. Fourteen operating mines in the southern field and the mountain district of Colorado had an aggregate capacity of 12,000 tons of coal daily. Four different plants, two in Las Animas and one each in Gunnison and Garfield counties, contained 800 ovens capable of producing 25,000 tons of coke monthly.[22]

Iron lands aggregating 2,311 acres were located in Colorado. The Orient and Calumet mines, with a maximum combined output of about 900 tons daily, supplied most of the ore consumed at the Bessemer Works. Among undeveloped properties, the most promising was the Ashcroft deposit in Pitkin County.[23]

The Colorado Fuel and Iron Company owned and operated the only integrated Bessemer plant in the West. Three blast furnaces had a combined capacity of 400 tons of foundry, pig, or ferro-manganese iron daily. The converter's two 5-ton vessels produced up to 500 tons of steel per day. The rolling mill, containing one three-high 35-inch blooming train and one three-high 23-inch rail train, had a maximum output of 500 tons of blooms and 300 tons of rails daily. The merchant mill's 20-inch muck train, 20-inch bar train, and 9-inch guide mill turned out 40 tons of merchant iron daily, plus a quantity of light rails. The cast pipe foundry yielded 40 tons, and the spike mill 5,000 pounds in a 12-hour shift.[24] Supporting units included a casting foundry, machine, carpenter, and pattern shops, and a laboratory.

It was assumed that with a relatively small expenditure for improvements, the steel works could manufacture iron, steel rails, and other products at costs that would assure competition in the territory west of the Missouri River. In addition, it was estimated that the demand would exceed the capacity of the Bessemer Works, a circumstance that seemed to guarantee operation of the plant at a maximum level of efficiency.[25]

The Colorado Fuel and Iron Company, in order to make the Bessemer Works an effective competitor in the steel industry, engaged experienced personnel to take charge of production facilities. Theodore Winthrop Robinson, a graduate of the Massachusetts Institute of Technology and supervisor of furnaces at the Milwaukee Works of the Illinois Steel Company, was named general superintendent of the iron and steel department. He remained in that position for seven years before returning to the middle-western corporation as general manager.[26]

Probably because of T. W. Robinson's influence, key subordinate positions were filled by men who had been associated with the Illinois Steel Company. Charles Snelling Robinson was given charge of the blast furnaces. Seven years later, he was promoted to fill the vacancy created by his brother's resignation.[27] Thomas J. Crowe came to the converter department from a similar position at the Joliet Works.[28] Thomas J. Brown, also of Joliet, was appointed head of the rail mill.[29] A. E. Jupp had been employed in the merchant mill at Milwaukee prior to assuming the leadership in the equivalent department at Bessemer.[30]

At the time of consolidation, it was recognized that at least $1,000,000 was needed for working capital and for improvements in the operating departments. Although the merger plan provided mortgage bonds for these purposes, the management decided that it was unwise to market them until the new corporation had demonstrated its earning power. Meanwhile, operations were sustained with time loans from Denver and New York banks.[31]

After only seven months, the fuel and iron company experienced a serious financial crisis. A slump in sales in the spring of 1893 undermined the corporation's ability to meet its short-term debts. In early May, the lack of funds caused the abandonment of all improvement programs at the steel works and mines.[32]

The crisis was magnified two months later by the Panic of 1893, triggered by the failures of the Philadelphia and Reading Railroad and the National Cordage Company. In July, after a number of bank failures had occurred in Pueblo and Denver, the fuel and iron enterprise encountered great difficulty in borrowing money for working capital. For a period of almost two years thereafter, the management struggled from month to month to avoid insolvency, never sure that it would survive the depression that stagnated the American economy.[33]

The Colorado Fuel and Iron Company was adversely affected by a combination of factors. The general business crisis, the decline of silver mining and the smelting industry, an unusually mild winter throughout the West, and strikes by coal miners and railroad workers restricted markets and disrupted production to the detriment of all departments.[34]

The Bessemer Works quickly felt the full impact of the Panic. The mills closed in July, ostensibly for repairs, and reopened only briefly in August because railroads, reacting to the financial crisis, cancelled orders for rails.[35] Almost all departments ceased operations by the end of the summer.

The steel plant remained idle until the next year. One blast furnace was started in February, another in May. The converter and rail mill resumed production in mid-March, but because of a shortage of pig iron, worked only part time until July, when they closed because of the coal strike.[36]

The coal miners' strike disrupted operations in the fuel department from late April until early July. All mines were affected, and a majority were closed at least part if not all of that time.[37] The strikers eventually returned to work without winning any concessions from the company.[38]

A nation-wide suspension of railroad traffic from June 29 to July 11, 1894, further disrupted production at mines and ovens and briefly halted shipments in and out of the steel works. The strike, called by Eugene V. Debs' American Railway Union, aroused some sympathy among miners and steelworkers. During the walkout, men from the general offices in Denver served as security guards at Bessemer.[39]

The company's financial condition improved in 1895, when a group of New York brokers agreed to market general mortgage bonds in the amount of $2,000,000. At the suggestion of Edward O. Wolcott and Dennis Sullivan, Moore and Schley Company offered to place the entire issue in return for a stock bonus of $1,750,000, or all of the undistributed common shares that had been authorized at the time of consolidation.[40] Osgood called the proposal exorbitant, but he did consent to give the brokerage firm one-half of the projected issue with a bonus of $375,000 in stock, plus an equal amount for expressing an interest in the transaction, on condition that he place the balance with other houses. The remainder of the bonds were assigned with a bonus of $375,000 plus an additional $50,000 to G. H. Prentice for special assistance in the matter. Therefore, the securities were sold at a cost of $1,174,000 in stock.

After the negotiations had been completed, Osgood had to find some way of carrying out the stock transfers without letting New York brokers know that Moore and Schley Company and Prentice had received extra bonuses, and at the same time to circumvent the New York Stock Exchange's unwritten rule against registering securities that had not been sold for cash, or equivalent, and actually distributed. Both problems were successfully resolved. The Exchange's Listing Committee, after some hesitation, accepted Osgood's word that the stock had been sold, after which the transactions were recorded in a manner that indicated that the company had realized from the sales of bonds and stocks the amount realized only from the former.[41]

The Bessemer Works returned to near-normal conditions in 1895, after the company joined with eastern manufacturers in establishing production quotas on steel rails. That year, the Colorado firm, in concert with others, was authorized to manufacture 60,000 tons, and subsequently received from the Rail Makers' Association a payment of $86,730 for participating in the pool. This raised receipts to $24.95 per ton, or more than one-third in excess of the actual selling price.[42]

The steel rail pool, called the "Rail Makers' Association," had been created in 1887 by the country's leading manufacturers for the purpose of fixing quotas on output and maintaining minimum prices on heavy sections. It was reorganized in 1893 and again in 1894, at which time the Colorado corporation joined the combination. The Illinois Steel Company withdrew that year but continued for many months a policy of cooperation with the monopoly.[43]

The Colorado Fuel and Iron Company signed on October 15, 1895, a second contract with the pool. The Pueblo firm agreed to charge in 1896 a minimum price of $31 for rails that weighed more than 45 pounds to the yard. In return, the association promised to maintain a comparable price in the East; guaranteed the western enterprise sales equal to 4 percent of the market, or 60,000 tons; and agreed to compensate the corporation for any deficiency at a rate of $5.75 per ton.[44]

One year later, the fuel and iron company agreed to halt all production of heavy rails. In return, the association promised 12 monthly payments of $30,000, commencing in February 1897.[45] The Illinois Steel Company, by terms of a corollary agreement, offered to manufacture for the western enterprise up to 10,000 tons of 16- and 40-pound rails at a price not to exceed $28 per ton.[46] This enabled the Colorado producer to supply its regional customers while work in the steel department was suspended at Bessemer.[47]

Price cutting by the Illinois Steel and the Lackawanna Iron and Steel companies caused the collapse of the pool in February 1897. Therefore, the eastern railmakers refused to honor the contract with the western manufacturer.[48] By then the rolling mill at Bessemer had been closed, and it was too late to solicit orders for that year. In 1898,

Osgood accepted $60,000 in settlement of all claims against the defunct association.

Following the dissolution of the rail pool, the price of heavy sections fell sharply, then rebounded to about $10 below the previous level, indicating that the larger producers imposed some restraint on the market. However, no attempt was made to revive the association until September 1898, when the formation of the Federal Steel Company, a consolidation of several enterprises, including Illinois Steel, upset the status quo within the industry.[49]

Negotiations dragged on for more than a month before the presidents of eight companies organized the Empire Rail Company as sole agent for the distribution of heavy steel rails in North America. Each participating corporation was assigned a quota of the market: Carnegie and Illinois Steel, 29.5 percent; Lackawanna Iron and Steel, 14 percent; Cambria Iron, 11 percent; Pennsylvania Steel, 9 percent; Colorado Fuel and Iron, 6 percent; and Maryland Steel, 1 percent. All rails were to be billed to the agent at the rate of $12 a ton, except that the western firm was allowed $13.75, and the proceeds were to be divided on the basis of the sales quotas. The pool was to be administered by Judge Elbert H. Gary of Federal Steel, Charles Schwab of Carnegie Steel, and Walter Scranton of Lackawanna Iron and Steel.[50]

That scheme was never implemented. Schwab announced on November 18, as the presidents gathered to sign the pooling agreement, that Carnegie Steel would not enter into any combination designed to restrict output and raise prices of steel rails. Osgood believed that the plan had been personally blocked by Andrew Carnegie. John W. Gates suggested that the industrialist's action was prompted by the fact that his firm could turn out T-rails at a cost that suggested it had more to gain from selling in a free market than by accepting less than one-third of all sales at a price controlled by a single agency.[51]

Acting on the assumption that Gates was correct, and that the two largest manufacturers would in fact divide the domestic market, Osgood sought an understanding with Schwab and Gary. The westerner's fears were justified, for the two men were negotiating an offensive and defensive alliance "with the idea of taking practically all of the business in the country."[52]

Osgood persuaded Schwab and Gary to include Cambria Iron and the Colorado Fuel and Iron companies in the pool. It was agreed that the two larger corporations were each entitled to 39.75 percent of all steel rail sales in North America. The smaller firms were to have 13.5 and 7 percent, respectively. Receipts in excess of $13 per ton, or $14.70 for the western member, were to be pooled and distributed on the basis of the sales quotas. This assured the Pueblo firm at least $19 dollars a ton and an output approximately equal to the capacity of its rail mill.[53]

The two giants of the industry admitted the smaller companies into the pool for selfish reasons. Carnegie Steel, according to Osgood, hoped to absorb the Cambria works; and the Federal Steel Company was not adverse to the idea of expanding operations by acquiring the Colorado enterprise.[54]

The understanding between the four rail manufacturers was implemented November 28, 1898, with the formation of the "New York Committee." That body handled all sales of sections heavier than 50 pounds per yard and distributed the proceeds to members. An office was opened in the Empire Building, adjoining Federal Steel's headquarters, and W. H. Vilas was employed as secretary and auditor. Each of the member companies appointed an agent to look after its affairs in the pool.[55]

The New York Committee, better known as the Rail Association, exercised centralized control over the distribution of heavy rails in North America for six years. Membership was expanded in 1899 to include Lackawanna Iron and Steel and Pennsylvania Steel. National Steel joined in 1900.[56] The pool successfully raised the minimum price for steel rails from $17 to $28 through the quota system. The Colorado Fuel and Iron Company was authorized in 1900 to sell to the limit of its rolling capacity, an arrangement that remained in effect until the combination was dissolved four years later.[57]

The Rail Association ushered in an era of prosperity for the Bessemer Works. The output of steel rails rose from slightly more than 17,000 tons in 1898 to more than 128,000 three years later, while receipts from that product advanced from $343,000 to more than $4,000,000 in the same period.[58] The return per ton of rails increased from $19.70 in 1898 to a high of $31.09 in 1901, and averaged $27.58 per year for the following two years.[59]

The western enterprise, despite its relatively small size among rail manufacturers, played an important role in the combination for a number of reasons. Its geographical location enabled it to influence a large regional market. Osgood, an aggressive businessman, was well known to eastern industrialists, who preferred to have him as a partner rather than as a rival. Perhaps more important, the men who controlled eastern mills also had close ties with railroads which were dependent on the Colorado enterprise for fuel. Finally, the Bessemer Works was expanded and modernized to make it a serious competitor in the West.

An improvement program was projected in 1896 for the purpose of lowering production costs and ending the company's reliance on the manufacture of rails. The blooming and rail mills were to be modernized and new products developed with the erection of plate and structural steel plants. That plan was abandoned early the next year when the collapse of the Rail Makers Association caused widespread dislocation in the market.[60]

The sharp decline in the price of rails focused attention on the necessity of ending old-fashioned production methods in the rolling mills. With that objective in mind, a half-million dollar program was authorized in the fall of 1897 to reconstruct the steel department. That work, financed with proceeds from a chattel mortgage, was completed the following summer.[61]

A second round of improvements was initiated in September 1899, when Osgood, in a letter to stockholders, requested support for an amendment to increase the corporation's capital from $13,000,000 to $25,000,000, the proceeds to be used in enlarging output and reducing costs in all operating departments. This was needed, he argued, if the company hoped to advance sales. Its natural marketing area, all of the United States west of the Missouri River, consumed approximately 1,500,000 tons of iron and steel products annually, but the Bessemer Works, at best, was able to supply only one-tenth of that amount.[62]

In a special meeting on September 25, 1899, stockholders unanimously approved the proposed increase in capital stock. More than one-half of the proceeds were tentatively allocated for new construction at Bessemer, and the balance was reserved for improvements in the fuel department and for working capital.[63]

That action climaxed a year-long struggle to underwrite a major expansion program. At first, Osgood had hoped to organize a new corporation. In fact, he contracted with Delos A. Chappell, Morgan Jones, and General Grenville M. Dodge, majority stockholders in the Victor Coal and Coke Company, for the purchase of that enterprise with the intention of consolidating it with the Pueblo firm to form the Colorado Steel Company with an authorized capital of $50,000,000. The plan was scuttled in July 1899, in part because New York brokers withdrew their support, and also because Osgood's personal finances were momentarily endangered by heavy losses from speculations in the stock and commodity markets.[64]

The purchase of the Victor company was completed that year. Operator of mines and ovens in Las Animas County, it had been a troublesome competitor for many years. Therefore, when the opportunity occurred, Osgood pressed for the acquisition with funds advanced by the fuel and iron enterprise.[65] Eventually, ownership of the rival concern passed to the Colorado Finance and Construction Company.

That corporation was created by Osgood and his associates to market the fuel and iron company's common shares. It was modeled after the construction firms that had been widely employed to build and equip railroads with the proceeds of their own stock. However, the Colorado promoters deliberately gave their "holding" company broad scope so that it could be used to finance and control a variety of business ventures.[66]

The Colorado Finance and Construction Company, organized October 4, 1899, was devised by Jerome to enable Osgood and his associates to engage in a wide range of activities. The Colorado corporation was empowered by its charter to acquire, control, and sell the securities of other enterprises; and to purchase and develop land, mines, ovens, plants, and other facilities required in the fuel and steel trades. The capital stock was limited to $100,000 so that it could be controlled by a small circle of investors. The directors for the first years were Osgood, Beaman, Cass, Jerome, and Samuel I. Heyn.[67]

The construction company, according to Jerome, was Osgood's personal business fief. He owned and ran the enterprise to suit his interests. The capital stock was not subscribed, stockholders existed only in name, and the directors convened only once to organize the corporation. Minutes of mythical directors' meetings were drafted by Jerome, the treasurer, and filed by Osgood, the president. He kept all records in New York and refused to make an accounting of the firm's finances to other officers and directors.[68]

On October 4, 1899, the Colorado Finance and Construction Company made a contract with the Colorado Fuel and Iron Company for major improvements at Bessemer and in the mining camps. Compensation for that work was to be one-half of the recently authorized common stock.[69]

In this instance, the construction company served as an agency through which the fuel and iron shares were marketed. This indirect method was adopted because Colorado law permitted the issuance of stock only at par value, which, given the condition of the market, was out of the question at the time. Osgood arranged with a syndicate of New York brokers to take 60,000 shares. The net return, amounting to $3,600,000, was more than enough to pay for the authorized improvements and to give the Pueblo industry a cash return of $2,000,000.[70]

Improvements commenced in the fall of 1899. At Bessemer, two blast furnaces were rebuilt, the rail mill was expanded and modernized, a subsidiary railroad system was created to serve mines and mills, and the water system was enlarged. Elsewhere, an iron mine was opened at Sunrise, Wyoming, additional coal deposits were developed in the southern field and in the mountain district, ovens were erected at Redstone and Sopris, washeries were installed at El Moro and Starkville, and an electric power plant was built at Starkville.[71]

The construction company had little to do with the projects other than to receive stock and to make periodic payments to the fuel and iron corporation. That enterprise planned all improvements, negotiated with contractors, and supervised actual construction. It made payments, charging disbursements to Osgood's ancillary firm.[72]

At first the promoter was disappointed with the construction company's attempt to fund the improvements. In placing the initial installments of the 60,000 shares, the syndicate received only 50 percent of par value, substantially less than anticipated. Even with an advance in the market, Osgood predicted there would be no profit in the venture.[73]

Seeking more lucrative investments, Osgood turned in early 1900 to the New Mexico fuel trade. From M. W. Flournoy, of Albuquerque, he purchased at a cost of about $300,000 the Crescent Coal Company, which operated mines at Gallup. With that asset, the Colorado promoter planned to raise half a million dollars, enough to pay for the property plus funds for development. A corporation was to be organized with a capital of $2,000,000, one-quarter in preferred stock. One share of the preferred and two of common were to be given for each subscription of $100, and the balance, equal to $500,000 of the latter, was to be divided between Osgood, Kebler, Cass, and Jerome. The venture had the potential of earning not less than 6 percent annually for the common shareowners.[74]

Similar schemes were devised for exploiting coal resources elsewhere in the West. Osgood was then deeply involved in negotiations with the Union Pacific Railway for the acquisition of its fuel deposits in Wyoming. Simultaneously, he was talking to the owners of the Northern Coal and Coke Company about the possibility of purchasing its extensive properties in Colorado. Both goals failed to materialize.[75]

Also, Osgood had decided that the Victor Coal and Coke Company should not be surrendered to the Colorado Fuel and Iron Company. Since there was no longer any hope of promoting the Colorado Steel Company, it seemed only proper that he and his associates should keep the enterprise for themselves.[76]

All business ventures related to the fuel trade were handled by the Colorado Finance and Construction Company. It purchased properties and solicited subscriptions for new enterprises. Osgood anticipated promoter's profits from all transactions in addition to the earnings of corporation, which he alone controlled. Furthermore, he was assured of a dominant influence in all firms organized in this way.

In March 1900, Osgood decided to create the American Fuel Company under the laws of New Mexico Territory in order to operate mines at Gallup. The enterprise was capitalized at $2,000,000, one-fourth preferred.[77] The next month, the contract with Flournoy was assigned to the construction firm, which organized the new venture. Subscriptions were solicited mainly among Denver capitalists, with $500,000 in common stock reserved as promoter's profits.[78] One-half of that amount was allocated to Osgood; the balance was divided among Kebler, Cass, and Jerome.

A similar venture was organized in southern Colorado. The Victor Coal and Coke, Colorado Coke, Gray Creek Coal and Coking, and the Gray Creek Coke companies were consolidated as the Victor Fuel Company with assets that comprised 14,000 acres, five mines, and 200 ovens in Huerfano and Las Animas counties. That enterprise's mortgage bonds, amounting to $1,000,000, were marketed by the Colorado Finance and Construction Company with the understanding that $950,000 was to be allocated for purchasing the properties, and $150,000 to Osgood as a commission on the transaction.[79] Each subscriber to a $1,000 bond at 110 received as a bonus an equal amount of six percent preferred stock. In this way, all of the common stock remained in the hands of the promoter, and with it control over the new firm's business affairs.[80]

Although deeply involved in mining ventures, Osgood did not overlook the opportunity for investments in connection with improvements at the Bessemer Works. In August 1899, Jerome suggested that Osgood's friends organize a real estate company to underwrite the construction of dwellings and other facilities adjacent to the steel works. The community, Jerome insisted, was "dirty, in poor repair, and repulsive," but with a modest investment could be made "comparatively attractive."[81]

This concern may have been inspired at least in part by the fact that the Colorado Fuel and Iron Company had large real estate interests in Bessemer, having reacquired much of the property that had been surrendered to the Colorado Coal and Iron Development Company at the time of consolidation. That corporation had been forced into bankruptcy in 1896, and Jerome, as the court-appointed receiver, had presided over the final liquidation of assets three years later.[82] The lands that reverted to the operator of the steel works were placed under the administration of the Pueblo Realty Trust Company, a subsidiary created in 1900.[83]

Under the court order that distributed the remaining assets, the fuel and iron company did not acquire all of the real estate that it needed for improvements inaugurated in 1899 at the steel plant. Prior to bankruptcy, the development corporation, in an attempt to eliminate its floating debt, had conveyed some of its property to the Suburban Land and Investment Company.[84] Included in that transaction was a narrow strip lying between the blast furnaces and Evans Avenue.[85]

That tract and adjacent acreage was purchased by the Colorado Finance and Construction Company from the Suburban Land and Investment and the Pueblo Realty trust companies. Except for the property immediately adjoining the steel works, it was conveyed in turn to the Minnequa Town Company, a new corporation, for a majority of its capital stock.[86]

The town company was organized May 29, 1899, under the laws of Colorado, to carry out improvements at Bessemer. One-quarter of the authorized $600,000 capital was in six percent preferred stock. Common shares in the amount of $150,000 were set aside as promoter's profits, one-half of which were assigned to Osgood and the balance to Kebler, Cass, and Jerome. [87]

The Colorado Finance and Construction Company, as principal stockholder, dictated the new corporation's selection of officers. George W. Bowen, who was also president of the American Fuel venture, was elected president, assisted by Kebler as vice-president, Jerome as treasurer, Heyn as secretary, and George Herrington as general manager. Several local residents were named to the board of directors. In addition to the officers, that body was composed of J. B. McKennan, Thomas J. Brown, Thomas J. Crowe, Howard A. Case, Fred Darrock, and Dr. Richard W. Corwin, all of whom held supervisory or comparable positions in connection with the steel works.[88]

The town company devoted its energies to the development of a tract that extended westward from the Bessemer Works. A portion was subdivided as Minnequa Heights, a residential area in which the corporation planned to build family dwellings for resale. Twenty acres was reserved as a site for a new hospital to serve the fuel and iron company's employees, and an amusement park was projected for the area at the north end of Lake Minnequa.[89]

The Colorado Finance and Construction Company had larger interests at stake than the Bessemer improvements. On April 4, 1901, it secured the last of the fuel and iron common stock that had been authorized for two years for expansion of the steel works. Sixty thousand shares were exchanged for $3,480,000 in cash and the outstanding capital stock of the Rocky Mountain Coal and Iron Company.[90] That corporation owned potentially rich coal lands in Las Animas County, comprising 258,000 acres of what was the northern portion of the Maxwell Grant.[91]

Also on April 4, 1901, the Colorado Fuel and Iron Company purchased from Osgood's construction enterprise the Laramie Iron and Steel Company, operator of a rolling mill in Wyoming. In exchange for $150,000 in cash, the Pueblo firm acquired all of the capital stock, accounts, contracts, and business goodwill of the small but troublesome competitor in the iron trade.[92]

The Colorado Finance and Construction Company reaped large profits from its many promotional endeavors. The sale of fuel and iron common stock yielded earnings of from $3,000,000 to $4,000,000. Income from the Victor Fuel Company amounted to almost $1,000,000 in four years. Additional gains may have been secured from the American Fuel Company, speculations in mortgage bonds of the Grand River Coal and Coke Company, the sale of the Laramie Iron and Steel Company, the purchase of iron lands in San Bernardino County, California, and the organization of the Steel Wheel and Wagon Company, a Pueblo manufacturing enterprise founded in 1901.[93] More important, the corporation enabled the fuel and iron company to complete needed improvements at the Bessemer Works.

A new phase of the reconstruction program was announced in May 1900. On a portion of the tract purchased from the Suburban Land and Investment Company, an office building, medical dispensary, and laboratory were erected to take the place of structures that had to be torn down within the plant grounds to make room for another blast furnace. Eventually, the main gate was to be relocated from Northern Avenue to Canal Street, and in anticipation of that change, work was started on a viaduct over the railroad tracks to expedite the movement of people in and out of the steel works.[94]

Other projects authorized at that time were located in other parts of the city. A large reservoir was to be constructed three miles south of Lake Minnequa, and a new hospital was to be built to replace the existing structure, which was inadequate in size and incapable of further expansion. Also, the Colorado Supply Company announced that it planned to erect a large retail store and warehouse near the new office buildings on Canal Street.[95]

In 1901, the fuel and iron company inaugurated its most ambitious improvement program. The proceeds of a multimillion dollar issue of debenture bonds were devoted to the construction of new mills in an attempt to end the firm's traditional reliance on the production of steel rails.[96]

The decision to build plants for the fabrication of new products was forced on Osgood by a Chicago business associate. In the spring of 1901, John W. Gates, a flamboyant financier, began buying fuel and iron stock, sending the price on the New York market in an upward spiral. By the end of April, the newcomer, claiming a majority of the outstanding shares, let it be known that he and his friends hoped to see the Pueblo plant establish auxiliary manufactures for the production of tin plate, wire, and nails.[97]

A farm boy from Du Page County, Illinois, John Warne Gates, after an unsuccessful venture as a hardware merchant, had perceived the enormous potential of the market for barbed wire. In the decade of the eighties, he was instrumental in carrying out a series of consolidations that created a monopoly over the manufacture of barbed wire. Furthermore, his boldness and energy led in 1898 to the formation of the American Steel and Wire Company, a New Jersey corporation that dominated the wire trade. This achievement elevated Gates to a position of prominence among financial capitalists of the era.[98]

An inveterate gambler, Gates played poker as a diversion and, according to legend, nearly always won. His habits of the gaming table extended to the business world where he delighted in corporate manipulation. He bought and sold enterprises for high stakes, challenging the financial giants of that day. Andrew Carnegie thought that Gates was a fool, and J. P. Morgan called him uncouth and dishonest. To the public, he was "Bet a Million Gates," a plunger who was always ready to pursue money and power against great odds.

While Osgood denied that Gates controlled the voting stock of the fuel and iron company, he was willing to appease the Chicago financier by expanding the product line of the Pueblo steel works. On May 10, 1901, an agreement was concluded with John J. Mitchell, a business associate of Gates and president of the Illinois Trust and Savings Bank, for the underwriting of debenture bonds in the amount of $15,000,000 to finance the construction of new plants. John Lambert, senior partner in the New York brokerage firm of Blair and Company, agreed to market the securities.[99] The five percent debentures were convertible into stock at any time within five years and redeemable in ten.

In a special meeting on July 13, 1901, stockholders of the Colorado Fuel and Iron Company unanimously authorized an increase in capital stock from $25,000,000 to $40,000,000, and approved the issuance of $15,000,000 in debenture bonds for that purpose. Two-thirds of the proceeds was to be devoted to improvements at the Bessemer Works and to operating capital; the remainder was to be used to acquire properties such as the Victor Fuel, Northern Coal and Coke, and American Fuel companies.[100]

Five weeks later, the board of directors endorsed a $10,000,000 program for new production units. The character of the steel works was to be fundamentally altered with the erection of six 50-ton basic open-hearth furnaces, a 40-inch blooming mill, 12- and 24-inch continuous mills, a hoop and merchant mill, rod mill, plate mill, sheet mill, wire-drawing mill, and nail mill. Large sums were also committed to the construction of the Colorado and Wyoming Railway and the Crystal River Railroad, as well as for working capital.[101]

In less than a year, it was apparent that some if not all of the debentures reserved for the acquisition of properties would be needed to complete the auxiliary plants and other improvements at the steel works. Construction costs had exceeded estimates, partly because of delays in deliveries of machinery and also because of several disputes between workmen and the contractors who were erecting structures and installing equipment.[102]

In June 1902, the proceeds from the final installment of the debentures were allocated. More than one-half of the anticipated funds were devoted to completing projects in progress. The balance was earmarked for the purchase of land for plant expansion at Bessemer, the acquisition of iron deposits at Sunrise, Wyoming, and additions to the steel works' water system, coke ovens, and houses for employees.[103]

Three weeks after approval of the final

phase of construction, a financial crisis forced the fuel and iron company to retrench. Contracts for improvements, including the acquisition of property, were cancelled if they did not impair the functions of the operating departments.[104] Although the crisis persisted and eventually resulted in a change of management, all major projects, with the exception of some modifications in the water system, were completed by the summer of 1903.

From July 1, 1899, to June 30, 1903, the Colorado Fuel and Iron Company spent more than $24,000,000 for improvements. That compared with a total of only slightly more than $2,000,000 in the previous seven years.[105] Four-fifths of all monies appropriated in the later period were devoted to rebuilding, expanding, and modernizing the steel works. The total expenditure for that purpose was $19,569,735.28. The balance was used mainly to acquire properties and to develop mines, ovens, and quarries in order to supply the iron and steel department with a larger volume of coal, coke, ore, and limestone.

By 1903, the steel plant, redesignated the Minnequa Works,[106] bore little resemblance to the property the Colorado Fuel and Iron Company had acquired at the time of consolidation 11 years earlier. There were five blast furnaces, three of which had become operational in the past two years. A, D, and E, each with a capacity of more than 400 tons daily, augmented B and C, formerly Nos. 3 and 1, respectively.[107] Furnace F was then under construction.[108] The converter had been rebuilt and expanded and the rail mill enlarged and modernized.[109] Among older features, only the merchant iron mill and the pipe foundry were unchanged. The so-called "east mills" were entirely new.[110] The major elements of that complex were the open-hearth furnaces; the merchant steel department with large blooming and two continuous mills; twin hoop and cotton tie mills; and the mills for the manufacture of tin plate, sheets, rods, wire, and nails.[111]

A new system of reservoirs was also constructed because the expansion of iron and steel output made obsolete the company's reliance on Lake Minnequa for its water supply. Reservoirs No. 2 and 3, located south of Pueblo, and Sugar Loaf, on the Lake Fork of the Arkansas River near Leadville, assured ample water for all furnaces and mills. Initially, water from the latter source was diverted from the river to the Bessemer Ditch, then pumped into a standpipe within the steel works. This arrangement was abandoned in 1906, with completion of the Arkansas Conduit, which carried water from the river at Adobe a distance of 39 miles to the new storage basins south of the Minnequa Works. With this system, the volume from the Arkansas River was sufficient for normal consumption; water impounded from the St. Charles during times of heavy runoff was used only on a standby basis and in the winter months when the conduit was inoperative.[112]

Reconstruction of the steel works forced the Colorado Fuel and Iron Company to augment its sources of iron ore. Until that time, the blast furnaces had been sustained by the Calumet and Orient mines, located on the enterprise's lands, and by Leadville contractors who extracted high-grade ores from the precious metal mines of Breece and Fryer Hills. These deposits were incapable of expansion to meet new iron production schedules.

Calumet had supplied the steel works since 1881. The magnetite's iron content and impurities varied with the depth of the workings. At first, the percentage of metal had been high, and for that reason the ore had been used in large volume even though it contained objectionable amounts of silicon and sulphur. The former impurity was largely eliminated in the blast furnace; the latter was not and it imparted undesirable qualities to the iron and steel. Initially "heap roasting" was employed to remove sulphur; later the ore was treated in small reverberatory furnaces. Neither method was satisfactory. Eventually, the ore's iron content declined to the point where the impurities were unacceptable. The mine was closed in 1896, reopened three years later, and finally abandoned, after having shipped a total of 289,000 tons to Pueblo.[113]

The closing of the Calumet property placed a heavy burden on the Orient Mine. Limonite from that source contained only about 44 percent metallic iron, but the ore body was seemingly large and free of impurities. Therefore, it was used extensively throughout the decade of the nineties. After 1895, the output increased to more than 100,000 annually, reaching a peak of 195,000 in 1899. However, by that date, it was apparent that the ore body was nearly exhausted. Prior to 1903, shipments amounted to 1,222,000 tons. Some production continued until 1931, and the property was not abandoned until 1942.[114]

Convinced that a large steel plant required enormous reserves of ore, the Colorado Fuel and Iron Company, through its geologist, Richard C. Hills, systematically explored Colorado and the West. His work was so thorough that no deposits of any consequence were reported after 1904, the year that he terminated his investigations.[115]

Hills analyzed and evaluated all known deposits in Colorado. With the possible exception of Ashcroft in Pitkin County and Taylor River in neighboring Gunnison, ore bodies were either being worked, as at Calumet and Orient, or they were unexploitable. Rich bog ores of Park and Gunnison counties were too small to merit development. Conversely, iron-bearing carbonates that abounded in southern Colorado were deficient in metallic content. The magnetite of Fremont and Boulder counties contained too high a percentage of titanium. It was apparent that the fuel and iron company had to seek ore reserves in adjacent states and territories.[116]

Experience had demonstrated, as in the case of Pennsylvania, that it was economically feasible to transport iron ore to regions that abounded in coking coal. That lesson was repeated in the West. In making the decision to expand the steel works in 1899, and again in 1901, Osgood and his associates knew that they would have to find ore reserves outside Colorado and, because of Hills' investigations, they knew where to seek them in New Mexico Territory, Wyoming, and Utah.[117] Furthermore, they recognized that the region's rail network afforded reliable and cheap transportation for moving bulk shipments of ore from distant mines to Pueblo. The first task, therefore, was to gain control of the sources of raw material.

Iron ore from the Hanover district of New Mexico, the first from outside the state, had been tested at the Bessemer Works in 1894. The area near Fierro had been opened in 1891 by the Southwestern Coal and Iron Company, whose sample of magnetite was high in metallic content and suitable for making Bessemer steel. However, no attempt was made at that time to obtain a regular supply, presumably because of the cost.[118]

Five years later, the Colorado Fuel and Iron Company decided to augment ore from Orient with a large supply from Fierro. On February 24, 1899, the Hanover-Bessemer Iron Ore Association leased 445 acres of iron claims to the Pueblo firm for a period of twenty years.[119] Three mines supplied a large tonnage of ore. The "86" and Jim Fair at Fierro commenced shipments in 1899, and the Union Hill, located midway between that community and Hanover, started two years later. The products were magnetite and hematite.[120] Additional ore was procured from I. W. Bible, formerly of Colorado Springs, as a by-product of his copper, silver, and lead enterprises.[121]

Rich deposits of iron ore were also known to exist near Hartville, Wyoming, 100 miles north of Cheyenne, but were generally ignored until 1887, when I. S. Bartlett filed claims in that area. An official of the Wyoming Copper Company, he had engaged for a time in mining and smelting copper ore. When that venture proved unprofitable, he attempted to develop the more abundant iron resources of the region. Through his efforts, a group of eastern capitalists, organized as the Wyoming Railway and Iron Company, purchased in 1890 the so-called "Sunrise Group" of claims in anticipation of erecting iron-reduction works.[122]

That year, the Colorado Coal and Iron Company showed an interest in the Wyoming iron lands. After a careful investigation of the Sunrise and Village Belle claims, George S. Ramsay concluded that they did not contain sufficient ore to justify commercial development. An estimated 500,000 tons of hematite was too small a quantity to justify construction of a railroad and the expense of opening and equipping a mine.[123]

Three years later, the Colorado Fuel and

Iron Company ordered a careful study of the deposits. In April 1893, Hills made the first of many trips to Hartville for the purpose of evaluating the ore claimed by various parties in the area.[124] Three years later, he contradicted Ramsay's earlier judgment, arguing that his predecessor had grossly underestimated the size and value of the ore in the Sunrise Group. He also noted that the Whalen Canon Group, of which the Chicago claim was partially explored, contained large quantities of high-grade hematite. Furthermore, Hills suggested, it was economically feasible to open the area for commercial mining.[125]

On November 5, 1898, the fuel and iron company leased the Sunrise Group, comprising 71 claims, for a period of 20 years. The contract with the Wyoming Railway and Iron Company provided that the lessee was to pay a royalty of $.15 for each ton shipped to the steel works at Pueblo and twice that amount for ore extracted for resale. The minimum output was to increase from 150,000 to 300,000 tons over a period of seven years. Also, the Colorado firm was to use its best efforts to gain construction of a railroad to the deposits within one year and, if unable to do so, had the option of terminating the agreement.[126]

The Sunrise Mine, an excavation worked by steam shovels, yielded a large output at a relatively small cost. The expense of mining and loading ore was about $.15 a ton. By the summer of 1900, 65 cars, or 2,000 gross tons, were shipped daily. Receipts at Pueblo jumped from 19,225 to 247,163 tons in three years.[127]

The Colorado Fuel and Iron Company built a railroad from the iron mine to connect with regional carriers having access to Pueblo. At first, ore was freighted to Porter, where it was forwarded by rail.[128] The Colorado and Wyoming Railway, a wholly-owned subsidiary of the steel corporation, completed in 1900 a standard-gauge line to the Cheyenne and Northern (later the Colorado and Southern Railway) at Hartville Junction. Subsequently, a branch to Guernsey provided access to the Chicago, Burlington and Quincy.[129]

Eventually, the Colorado company purchased most of the claims in the Platte Canon Mining District near Hartville. David Wegg sold his majority interest in the Wyoming Railway and Iron Company in 1902 for $500,000.[130] The titles were blocked by litigation until 1904. Six years later, the balance of the workable deposits were obtained from the Hartville Iron Company.[131]

Prior to 1903, additional iron deposits were acquired in Iron County, Utah. By purchase and location on the public domain, the Colorado company brought under its control hundreds of acres of rich hematite and magnetite. Many years later, these claims, particularly the Desert Mound lands near Cedar City, became an important source of ore for the Pueblo steel works.[132]

Another Utah property, the Little Grand Manganese Mine, was purchased in 1902 from the Colorado-Grand Development Company. That enterprise had been organized the previous year by David C. Dodge, Edwin Thatcher Wolverton, and others to exploit a seemingly large and valuable body of ferromanganese iron located in Wheeler's Desert about midway between Green River and the Utah-Colorado border. In the hope of gaining its own supply of raw material for the manufacture of spiegeleisen, the Colorado Fuel and Iron Company bought 25 claims, containing an estimated 300,000 to 400,000 tons of ore at a cost of $86,000.[133]

The mine was unique. Violet-black rocks and boulders were gathered from the desert, crushed, and transported 16 miles over a "turnpike" to Little Grand Station on the Denver and Rio Grande Western for shipment to Pueblo. A 17-ton traction engine pulled a water tank and two or more Studebaker wagons, each loaded with from 10 to 20 tons of ore at a maximum speed of 4 miles an hour on the trip to the railroad. Supplies of water were spotted along the route to sustain the steam engine which consumed as much as 100 gallons an hour in making the passage.[134]

The mine was closed after only two years because the cost was prohibitive. The Pueblo company invested about $130,000 in the land, equipment, and operations, and extracted from the property only 481 tons of ferro-manganese ore.[135] It was more economical to purchase that material from precious metal mines at Leadville or, if that source failed, to import spiegeleisen from Europe.

By purchase and lease, the Colorado Fuel and Iron Company secured ample iron ore to sustain the enlarged steel works. Prior to 1899, known accessible deposits in Colorado contained an estimated 4,000,000 tons of ore, three-fourths of which was in the largely unexplored Ashcroft lands.[136] Four years later, the corporation's reserves at Hartville and Fierro exceeded by conservative estimates 5,000,000 tons. In addition, the major deposits of Utah, of which the Pueblo enterprise owned a large share, surpassed 100,000,000 tons.

Limestone, unlike iron ore, was found in ample quantities close to Pueblo. The San Carlos Quarry, located at Lime six miles south of the steel plant, had been a major source of fluxing material since 1881. From that date until 1897, the property was leased to contractors who mined, crushed, and loaded the stone on railroad cars at a specified rate per ton. In the latter year, the fuel and iron company assumed that work, using its own employees as day laborers.[137]

Limestone was blasted from the quarry bed, broken with sledges into pieces convenient for loading into pit cars, and transported to the crusher. There it was reduced to a standard size about two inches in diameter, after which it was screened to remove shale and placed on railroad cars. Shipments to the steel plant averaged about 600 tons daily, and a smaller amount was sold to local smelters of precious metals.[138]

A second source of limestone was opened in the latter part of 1903 at Howard, later called Calcite, 50 miles west of Canon City in the Sangre de Cristo range. This new source of fluxing material enabled the steel works to manufacture low-phosphorous steel rails which were in demand by railroads at the time.[139]

Four quarries employed as many as 200 men and shipped up to 15,000 tons of fluxing and refractory materials monthly. Numbers 1 and 2 produced a salt-water-deposit limestone called "blue stone" used in the smelting of iron destined for the converter and rail mill. Number 3 yielded magnesium limestone, or dolomite, required for linings of basic furnaces. Number 4 produced a fresh-water stone which because of its purity was consumed in the manufacture of open-hearth steel.[140]

With the acquisition of iron and limestone properties and their development, the Colorado Fuel and Iron Company owned and operated its own mines and quarries to provide essential raw materials for the production of basic metals. However, the critical ingredient, in the Pueblo firm's experience, for making iron was coke. In past years, the inferior quality of that commodity had undermined the efficiency of the blast furnaces. The corporation's reconstruction program called for large-scale output of pig iron. The achievement of that goal demanded a high-grade coke. For that reason, the expansion of the fuel department was a key factor in the revitalization of the Minnequa Works.

Pueblo Plant

Tin plate mill construction, Pueblo, August, 1903.

Construction of new spike mill, Pueblo, August, 1903.

Pueblo plant construction, August, 1903.

Furnace Row at the Minnequa Works, Pueblo, ca. 1903.

10-inch mill, Minnequa Works, Pueblo, 1910.

Construction of billet yard, Pueblo, August, 1903.

Dockey engine, slag hopper, and crew. This apparatus was employed to transport waste from the blast furnaces to the slag pile that stretched southward from the plant site. Courtesy of The State Historical Society of Colorado.

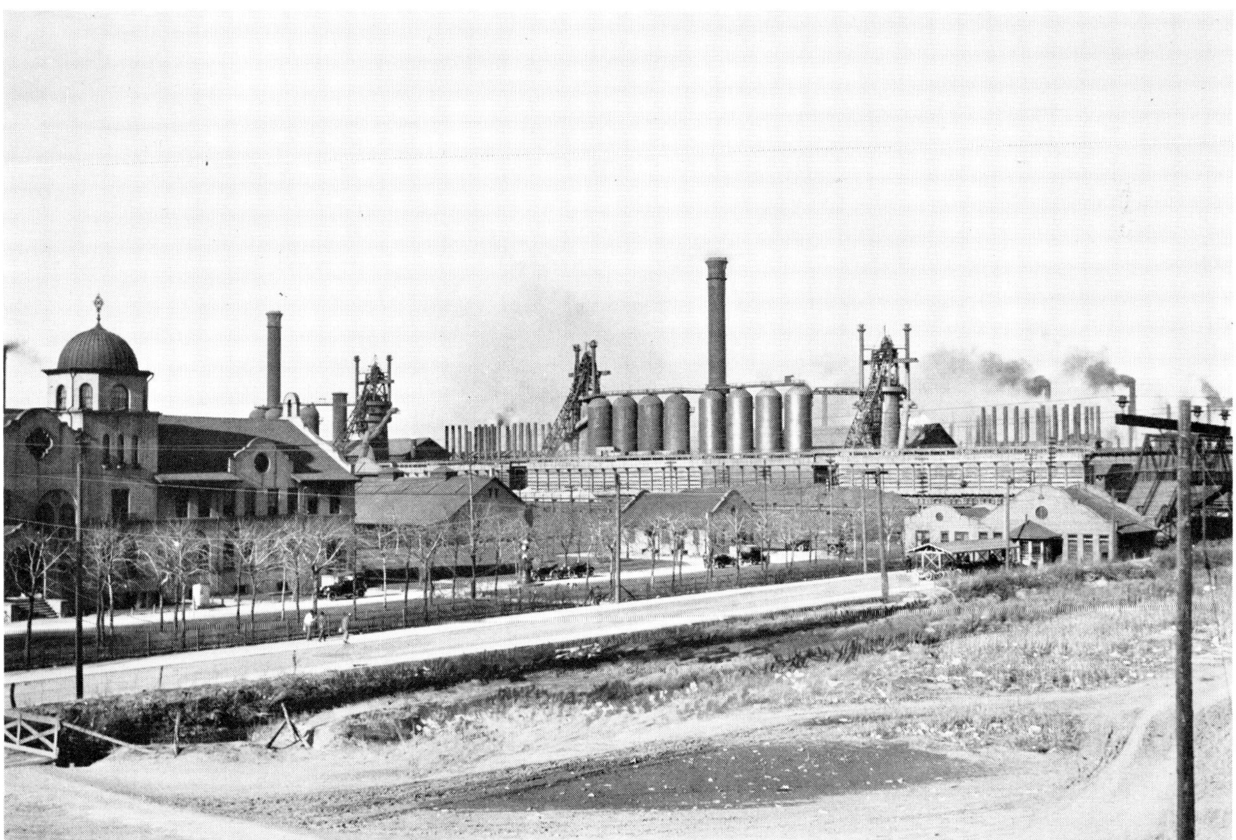

The Colorado Fuel and Iron Co.'s Minnequa Works, ca. 1916. At the right, flanked by the Bessemer canal, is the entrance to the plant. The dispensary is in the center, outside the plant confines, and the administrative building is at the left. Courtesy of The State Historical Society of Colorado.

The Colorado Supply Co.'s store at the steel works in Pueblo, ca. 1915.

W. H. Loughru in The Colorado Supply Co.'s delivery truck, Pueblo, Nov. 1915.

Federal truck operated by The Colorado Supply Co.'s wholesale grocery department in Pueblo, ca. 1916.

The Young Men's Christian Association took over control in 1915 of all club houses and social activities previously administered by the CF&I Co.'s sociological department. The "Steel Works Y" became a central feature of John D. Rockefeller's welfare capitalism scheme that was widely copied by American enterprises after World War I.

11-inch Precision Bar Mill, opened at Pueblo on June 16, 1971. Nearly two years in construction, at a cost of $41 million, floor space on the two-tiered structure is equal to about eight acres, or the equivalent of six football fields.

Basic Oxygen Furnace, Pueblo Plant.

Initial tap of electric furnace, Pueblo Plant, Oct. 22, 1973.

Main pulpit from which all 16 stands of CF&I Steel Corporation's 11-inch Precision Bar Mill can be controlled.

"The Patriot," a Colorado and Wyoming Railway locomotive dressed up to celebrate Colorado's centennial and the Nation's bicentennial in 1976. The carrier is a wholly-owned subsidiary which serves not only the Pueblo Plant, but links the Allen Mine west of Trinidad and the Sunrise Mine in Wyoming with mainline carriers that transport coal and iron ore to Pueblo.

A Colorado and Wyoming Railway locomotive moves a string of gondolas through a loading silo at the Allen Mine, west of Trinidad, as a unit train prepares to transport coal from Colorado's southern field to Pueblo for consumption at the steel works. Since the closing of Morley in 1956 and Frederick in 1960, the Allen Mine, opened in 1951, is the CF&I Steel Corporation's only coal producing property.

Rail Mill loading dock, Pueblo Plant.

Furnace Row, Pueblo Plant, 1974. Slightly to the left of center in this picture is a road sign on Interstate 25, built in the 1960's, separating the CF&I Steel Corporation's office complex from the plant.

Limestone Quarries

CF&I Co. office at Calcite Quarry, 40 miles west of Canon City. This source of limestone and dolomite was opened in 1904 to supplement fluxing materials quarried at San Carlos. The new quarry was located at Howard, later renamed Calcite. That property was abandoned in 1930, when another quarry was opened at Monarch.

Employees' club house at Calcite, 1918.

The Colorado Supply Co. store at Calcite, Jan. 4, 1918.

Limestone crushers at San Carlos Quarry, six miles south of the Steel works. The property, also known as Lime, was opened in 1881 to supply fluxing material for the Colorado Coal and Iron Co.'s blast furnace. San Carlos continued to supply the steel works with limestone until operations were suspended during the Great Depression.

The Colorado Supply Co.'s store at San Carlos, or Lime, ca. 1915.

Sunrise Iron Mine

Church at Sunrise, Wyo.

School house and pupils at Sunrise, Wyo., ca. 1915.

Sunrise Iron Mine and the community of Sunrise, Wyo. The property was leased by the Colorado Fuel and Iron Co. in 1898, and purchased in 1904. Ore shipments to the Minnequa Plant commenced in 1901. The mine has remained a major source of iron ore for the Pueblo steel works for 75 years. In the background is the open pit from which ore was taken until 1940. The shaft to underground workings is to the right of the pit.

Club house, Sunrise, Wyo.

Interior of employee's clubhouse at Sunrise, Wyo., ca. 1915.

Colorado and Wyoming Railway depot, Sunrise, Wyo.

Sunrise Mine boiler house and shaft for underground workings, ca. 1915.

Colorado Supply Co. store, Sunrise, Wyo., ca. 1915.

9

Expansion of the Fuel Trade, 1892-1903

For nearly a decade, the Colorado Fuel and Iron Company, following the example of its predecessors, directed most of its energies to the fuel trade. Additional mines were opened in the southern field and the mountain district, while throughout the state the corporation added to its already substantial holdings of coal deposits. Rival firms were absorbed or controlled by purchase or lease, strengthening the Pueblo enterprise's already formidable position in the western market.

The fuel and iron company began operations on October 21, 1892, as the region's foremost producer of coal and coke. It had acquired 14 mines from predecessors: Engleville, Sopris, and Berwind in the Trinidad district; Walsen, Robinson, Rouse, and Pictou near Walsenburg; Crested Butte and Anthracite in Gunnison County; Marion, Sunlight, and Newcastle in Garfield County; and Spring Gulch in Pitkin County.[1] Coke plants were maintained in conjunction with mines at Trinidad, Crested Butte, and at Cardiff, south of Glenwood Springs.

The corporation also owned nearly 70,000 acres of coal land. More than one-half of this amount was in the southern field, embracing Fremont, Huerfano, and Las Animas counties. About one-third of the fuel resources were in the mountain district, encompassing Garfield, Gunnison, Mesa, and Pitkin counties. A small tract was located in the lignite field north of Denver, and about 8,000 acres were in Carbon County, Wyoming. Altogether, these properties contained an estimated 400,000,000 tons of lignite, subbituminous, bituminous, and anthracite, which made ample reserves for the company's future needs at the steel works and for the commercial market.[2]

The company decided in 1892 to expand the production of anthracite and coking coal in the mountain district. The first project was a new mine at Ruby, better known as Floresta, where a drift was opened as a replacement for the Anthracite Mine. The latter property, it was anticipated, would be exhausted in five years. In fact, production continued there for more than a decade.[3]

Located in a narrow canyon on the north slope of Ohio Mountain at an altitude of 10,000 feet, Floresta was the highest coalmining camp in the United States, if not the world. The product, comparable to the finest Pennsylvania anthracite, was marketed in Colorado and throughout the West.[4]

Although as many as 94 miners could be employed at Floresta, they were usually in short supply and the property normally yielded less than capacity. A large percentage of the workmen in the camp at any given time were bachelors. Isolation and the fact that the mine was closed at least half of each year because heavy snowfalls blocked the Denver and Rio Grande narrow gauge branch from Crested Butte discouraged workers, especially those with families, from going there. Furthermore, a thin, 3-foot seam with a steep angle of inclination made it impossible for inexperienced miners to earn acceptable wages extracting coal.[5]

Farther north beyond McClure Pass, Osgood and his associates proposed to open the vast fuel resources of Coal Basin and to manufacture coke on the site. In carrying out this scheme, they projected a transportation system linking the remote area with the tracks of the Denver and Rio Grande and the Colorado Midland railroads in the Roaring Fork Valley. The Crystal River Railway Company was organized August 8, 1892, under the laws of Colorado.[6] That corporation purchased by exchange of stock the assets of the Crystal River Toll Road, the Colorado and Utah Railway, and the Aspen and Western Railway.[7] A new road was surveyed and graded from Carbondale to Coal Creek. In

January 1893, the Crystal River Land and Improvement Company, another of Osgood's many business ventures, agreed to underwrite the construction of a narrow-gauge line from the Crystal River to Coal Basin.[8]

At the latter place, eight miles west of the future site of Redstone, the Colorado Fuel and Iron Company planned to open a mine to produce an excellent grade of coking coal. The deposit comprised four seams aggregating from 20 to 30 feet of clean fuel which, as indicated by tests conducted at Sopris on a small sample two years previously, yielded coke superior to any then manufactured in the West. It closely resembled the celebrated New River product of West Virginia in that it did not readily break or crumble in handling. Characterized by a low percentage of ash and an absence of undesirable impurities, the coke possessed both toughness and porosity, qualities highly valued by smelters of precious metal ores.[9]

The coal and iron company's expansion program was interrupted by the Panic of 1893 and the ensuing depression. A prolonged period of financial stringency forced suspension of work at Coal Basin and on the railway in the valley of the Crystal River. Both projects remained dormant until the economic crisis passed.

The depression also forced Osgood and his friends to forego plans for additional mines and coke ovens in Las Animas County. On April 15, 1893, they had organized the Rocky Mountain Coal and Iron Company and on the same day negotiated with the Maxwell Land Grant Company a contract for the purchase of its property in Colorado at a cost of $1,000,000.[10]

By terms of that agreement, the Pueblo company was to acquire approximately 400 square miles of land lying west of Trinidad between the Purgatoire River and the Colorado-New Mexico boundary. Almost two-thirds of the tract was underlaid with bituminous coal, the possession of which gave the fuel and iron corporation virtually complete control over coking coal reserves in the southern portion of the state.[11]

However, the Panic of 1893 prevented the Rocky Mountain Coal and Iron Company from executing the agreement. When the initial installment of $200,000 fell due on July 1, 1893, the Colorado enterprise did not have the requisite funds. After numerous delays, the firm eventually forfeited its $10,000 advance, and the contract was canceled.[12]

The fuel and iron company's financial distress reached crisis proportions in 1894 when income from the fuel trade declined slightly more than half of previous levels. This condition was attributed to the general business depression and the decline of silver mining and the smelting industry in Colorado. Contributing factors included an unusually mild winter that reduced domestic coal consumption and a strike by Colorado coal miners.[13]

The labor dispute, lasting from late April to early July, had its origin at Coal Creek, spread throughout Fremont County, and eventually encompassed all of the southern field and most of the mountain district. The Colorado Fuel and Iron Company curtailed operations at the steel works and closed many of its mines and ovens. Two other large producers, the Trinidad Coal and Coking Company and the Victor Coal and Coke Company with properties in Las Animas and Huerfano counties, were similarly affected.[14]

The cause of the strike was in itself a matter of controversy. Osgood, Kebler, and spokesmen for other operators insisted that it had been called by the United Mine Workers of America in support of a walkout by miners in protest of wage cuts in the East. Since the Colorado companies had not altered wage scales, the strike was labeled unjust and meaningless.[15] Representatives of the miners argued, however, that there were legitimate local issues. They noted that men had not been paid for two months or more, and what compensation they had received for their labor was in the form of scrip that could be exchanged only for merchandise at company stores.[16] Later, D. J. Reed, State Inspector of Coal Mines, reported that irregular payment and the scrip system had been "the mainsprings that precipitated the strike."[17]

The strikers employed a combination of new and old techniques in pressing their demands on management. The most important innovation was collective persuasion, or what the operators called group intimidation. This took the form of mass marches to mines which had not joined the walkout. In this way, demonstrators from Fremont County closed mines at Rouse, Walsen, and Gray Creek, and appreciably thinned the ranks of coal diggers even where work continued under the protection of armed guards.[18]

Strikers also resorted to the joint convention, a device that had been used in similar circumstances a decade earlier. They formally asked operators to meet with spokesmen for the miners to discuss grievances and to seek ways of ending the walkout. For this purpose, delegates from most of the southern mines, a few from the Denver district, and Blossburg and Madrid, New Mexico Territory, convened in Pueblo. However, the owners and operators refused to participate unless the miners first returned to work. After three days, the strikers' representatives issued a "manifesto" and adjourned.[19]

That pronouncement set forth the miners' demands. They asked for recognition of the United Mine Workers of America as bargaining agent for workers in coal mines throughout Colorado and northern New Mexico; an effective system of check weighmen in all mines; compensation for digging coal at a ton-rate based on two thousand pounds; semimonthly payment of wages in lawful money; abolition of scrip and the truck system; an end to discrimination against union members; restoration of the prestrike pay scale at Newcastle; and strict enforcement of state laws pertaining to operators' obligations in supplying miners with timber, rails, and other materials in underground working places.[20]

The "manifesto" was the strikers' final act in concert. The back-to-work movement was already under way; by early July, most mines had resumed normal operations. Strife continued only at Newcastle where the miners rejected a reduction in wages. The fuel and iron company, insisting that the previous scale had been unrealistically high, closed the mine until October, by which time the opposition had disappeared.[21]

The miners' complaint about the use of scrip was directed in part against the Colorado Supply Company. That enterprise had been organized in August 1888 by Osgood and his associates to sell general merchandise and to maintain boarding houses in mining camps.[22] The first store opened at Sopris in September of that year, followed by similar ventures at Rouse and Pictou. With the formation of the Colorado Fuel and Iron Company in 1892, retail operations expanded to Bessemer and to coal communities previously controlled by the Colorado Coal and Iron Company.

The fuel and iron company, in cooperation with the supply company, adopted scrip as a method of providing credit for employees. A mill worker or miner who had accrued time on the payroll could receive on request an advance, or acceptance, equal to money actually earned, redeemable at a supply store. To facilitate these transactions, the employer issued the supply firm's scrip in fractional amounts, plus denominations of from one to five dollars, exchangeable for goods at a purchase price at par with coin.[23]

Following the Panic of 1893, the Colorado Fuel and Iron Company, in the absence of cash with which to pay employees, offered scrip. This was objectionable because the paper was redeemable only for merchandise at stores where, according to the strikers, prices were higher than those charged by independent merchants.[24]

As soon as the effects of the strike had passed and the fuel trade showed signs of improvement, the Pueblo enterprise expanded operations in the Canon district. The corporation leased in December 1895 800 acres along Chandler Creek, about 4 miles southeast of Canon City, from the Western Fuel Company, a small coal producing firm controlled by Pueblo businessmen.[25] The property was worked by two shafts, each about 500 feet in depth, which allowed access to a 5-foot seam of excellent domestic fuel. The mine was worked until September of the following year, yielding only a small tonnage.[26] The contract was not renewed, probably because the lessee had an opportunity to acquire more lucrative properties in the area and elsewhere in the southern field.

The Pueblo firm leased on August 20, 1896, all of the Colorado facilities of its principal competitor in the fuel trade, the Atchison, Topeka and Santa Fe Railway

Company. Involved were the assets of three subsidiaries: The Trinidad Coal and Coking Company, Canon City Coal Company, and Vulcan Fuel Company, operators of mines at Starkville near Trinidad, Rockvale and Brookside southeast of Canon City, and at Newcastle, respectively.[27] The fuel and iron company gained the exclusive right to mine coal from nearly 20,000 acres of land for which it agreed to pay a royalty of from $.06 to $.10 per ton. In this manner, the corporation eliminated its major rival in the fuel trade; at the same time, the railway secured not only a large income from its property but also profited from traffic in coal and coke from the lessee's mines and ovens.

As part of the arrangement with the Santa Fe Railway, the Colorado enterprise agreed to market fuel produced by the carrier's subsidiaries in New Mexico Territory. The fuel and iron corporation became the selling agent for the Cerrillos Coal Railroad Company, operator of mines at Madrid and ovens at nearby Waldo. On October 5, 1899, the Pueblo firm leased both properties for a period of ten years.[28] These agreements were not profitable, but the losses were insignificant compared to profits derived from the Colorado properties. More important, as long as the lease for the Cerrillos mines remained in effect, a potential source of competition was precluded from participation in the western fuel trade.[29]

In 1896, the Colorado Fuel and Iron Company gained almost complete control of the Canon district. In addition to the Rockvale and Brookside properties acquired by lease from the Santa Fe Railway, the Pueblo firm purchased the Fremont Mine from the Oak Creek Coal Company, a subsidiary of United Coal. Four years earlier, that enterprise had opened a shaft in Bear Gulch, a mile west of Williamsburg, to extract coal on land leased from the Santa Fe.[30] With this acquisition, all of the large producers in the area came under the direction of a single corporation.

No additional mines were developed until 1899, when one was opened and two were closed. The underground workings at Rouse flooded in the spring of that year, after which a replacement site was chosen five miles farther south where the Colorado Coal and Iron Company's Santa Clara camp had been located from 1887 to 1891. Buildings, machinery, houses, and personnel were transferred to the new community, which the residents called "New Rouse." In July, a slope was completed to the Robinson seam, the uppermost of three strata in the area, and shipments commenced immediately over a branch of the Denver and Rio Grande Railroad.[31]

Production also ended at Newcastle in 1899. The decision to close and dismantle the mine stemmed from two serious explosions and fires. The first occurred in 1897, when a defective fuse ignited gas, following which a fire swept through a portion of the underground workings. On that occasion, an irrigation ditch was turned into the shaft until the flames were quenched.[32] Two years later, another blast prompted permanent suspension of operations. For another decade, the property was worked under lease from the adjoining Coryell Mine.[33]

By 1899, development work had started on two coal deposits in the Crystal River Valley. That year, a drift was opened at Placita to furnish fuel for the Crystal River Railroad, which terminated at the mine, and to supply coke ovens under construction four miles to the south at the new town of Redstone. The production of bituminous coal ceased in 1901 after shipments from Coal Basin attained sufficient volume to fill the local demand.[34]

The Coal Basin Mine was opened by means of a rock tunnel 1,200 feet in length to three seams of coal, the third of which was clean and free of impurities. The underground workings were more than 300 feet above the tipple. Shipments began in December 1900, after the so-called "high line" was constructed from Redstone.[35]

The coal producers of Pitkin County were served by the Crystal River Railroad. That enterprise, incorporated September 6, 1898, was formed by a reorganization of the Crystal River Railway, in receivership. A standard-width line was completed in September 1899 from Carbondale to Placita, a distance of 21 miles.[36] A narrow-gauge branch extended from Redstone to Coal Basin the following year was an extraordinary feat of construction. It climbed 2,242 feet in 12 miles, featured a maximum grade of 4.4 percent, and a curvature up to 40 degrees. It was reported that a 13-car train had an opportunity to "straighten out" only once in traversing the rugged terrain.[37]

The exploitation of the coking coal of Pitkin County coincided with an attempt to open similar deposits in southern Colorado. An expansion program inaugurated at the Pueblo steel works in 1899 created the prospect of a severe fuel shortage, because the company's existing mines and ovens were incapable of supplying new furnaces and mills in addition to supplying the commercial market. Therefore, five new coal producers were activated in Huerfano and Las Animas counties in a period of four years.[38]

One mile northwest of Berwind, on the Ludlow Branch of the Colorado and Southern Railway, the company established the Tabasco Mine. A single slope provided access to a five-foot seam of coking and steam coal. Shipments commenced in April, 1901, while the major portion of the output was converted into coke in beehive ovens located on the premises.[39]

Farther north, the Hezron Mine commenced operation in September 1901. Located about midway between "Old Rouse" and "New Rouse," seven miles south of Walsenburg on a spur of the Colorado and Wyoming Railway, the camp's five slopes produced bituminous coal for railroad consumption and for the regional market.[40]

Mines were also opened in the Maxwell Grant south of the Purgatoire River. The Rocky Mountain Coal and Iron Company secured in 1901 the land for which it had bargained unsuccessfully eight years earlier. The property was assigned in turn to the Colorado Fuel and Iron Company, which immediately exploited the additional resources for the manufacture of coke.[41]

The first of the new mines in the Maxwell Grant was located 17 miles west of Trinidad. Initially called "Smith Canon," the community's name was changed to "Purgatoire," and in 1902 to "Primero." Six openings, all equipped with electric and steam haulage systems, had a combined capacity of 4,000 tons daily, making this the largest of the Pueblo firm's mines. All slack and screenings, constituting more than one-half of the output, was sent to coke ovens at nearby Segundo.[42]

Thirty-two miles west of Trinidad, work commenced in 1902 on another mine within the limits of the Maxwell Grant. Originally called "Torres," then "Rincon," and finally "Tercio," the camp was located in the Stonewall district in the foothills of the Sangre de Cristo Range. Six drifts yielded 1,000 tons of coal daily from seams that varied in thickness from 4 to 12 feet. More than four-fifths of the produce was assigned to coke ovens located near the mine.[43]

Cuarto Mine was located at the far end of Tercio Park. The work of driving four drifts commenced in November 1902, and coal was shipped four months later. Slack and screenings were sent to the ovens at Tercio, and all lump grades were sold to the Santa Fe Railway. Although an excellent steam coal, the product was deficient in coking qualities.[44]

All the new mines and ovens within the limits of the Maxwell Grant were served by the Colorado and Wyoming Railway, a wholly-owned subsidiary that had been organized in 1899 to link the iron mine at Sunrise, Wyoming, with nearby railroads over which ore was shipped to Pueblo. Another line that was constructed in 1901 served the mining camps of the Purgatoire Valley with regular passenger and freight service. Initially the track was built from Trinidad to Primero and Segundo; later it was extended to Tercio and Cuarto.[45]

By terms of an agreement effective March 1, 1903, the Colorado and Wyoming Railway gained the exclusive right to transport over its line at specified rates all traffic to and from certain of the Colorado Fuel and Iron Company's properties. Also, the subsidiary was guaranteed net earnings of $25,000 per month, sufficient to pay all charges on mortgage bonds in the amount of $4,500,000, which were surrendered to the parent enterprise in repayment of monies it had advanced in the previous four years to construct and equip four operating units.[46]

The Northern Division ran from Sunrise to Hartville Junction, with a branch to Gurnsey, connecting with the Colorado and

Southern and the Burlington and Missouri, respectively. In February 1903, the two lines handled about 25,000 tons of ore, an amount that more than doubled when all the new furnaces and mills became operational at the Minnequa Works.[47]

The Middle Division, within and adjacent to the steel plant, connected with all railroads that entered Pueblo. Colorado and Wyoming switch engines handled all shipments in and out of the Minnequa Works, amounting to more than 12,000 cars a month, plus all special equipment such as ladle, ingot, and slag cars. While Reservoirs No. 2 and No. 3 were under construction, the carrier hauled slag from the furnaces for use in lining the sides and bottoms of the storage basins.[48]

The Southern Division linked the Santa Fe's main track at Jensen with mines and ovens west of Trinidad. Over a leased line, the Colorado and Wyoming also entered Sopris where it connected with the Denver and Rio Grande and the Colorado and Southern railroads. Shipments from Primero and Tercio in 1903 averaged about 85,000 tons per month, and the new mine at Cuarto, along with ovens at Segundo and Tercio, contributed a large volume to intraline and interline traffic.[49] In Huerfano County, a short line connected the Hezron Mine with the Santa Clara Branch of the Denver and Rio Grande Railroad.

The subsidiary carrier, by giving the Pueblo corporation access to the region's railroad network, usually two or more competing lines, immeasurably contributed to the expansion of fuel operations. The output of mines nearly tripled from 1893 to 1903. The 3,750,000 tons recorded in the latter year would have been larger had not a coal miners' strike adversely affected production in November and December. The company's share of the state's total yield increased from 44 to 49 percent in ten years.[50]

Las Animas County was the principal center of coal production, followed in order by Huerfano, Fremont, Gunnison, Pitkin, and Garfield. In 1893, the Trinidad district had accounted for 42 percent of the company's output, and ten years later, 52 percent. In the same period, the Walsenburg area's proportion declined from 30 to 20 percent, and the Canon district's share rose from 7 to 15 percent.[51]

In addition to the products of its own mines and ovens, the Colorado enterprise marketed the output of competitors, particularly those in New Mexico Territory. This practice stabilized prices in the fuel trade. For several years, the fuel and iron company acted as selling agent for the Raton Coal and Coke Company, a Santa Fe Railway subsidiary. The Pueblo firm also purchased coal and coke from the Blossburg Coal Company, and continued to do so long after the expiration of a formal agreement for that purpose.[52] Similar relations were maintained with other producers throughout the region.

From 1900 to 1902, the Pueblo firm administered the properties of the American Fuel Company. That enterprise had been formed by Osgood and his friends from the Cresent Coal Company, whose properties were located in the Gallup district. In addition to working the Catalpa, Gallup, and Weaver mines, the fuel and iron corporation erected in that vicinity an experimental plant to manufacture briquettes. Refuse lignite was mixed with tar and pressed into bricks which, it was anticipated, would be as serviceable as anthracite for domestic and commercial consumption. About 400 tons of slack were used daily in this venture.[53]

The Colorado Fuel and Iron Company was the region's largest manufacturer of coke. In 1903, its nine sets of ovens, aggregating 2,971 units, produced 541,112 tons, or about 70 percent of the total in Colorado.[54] The only competitors were the Victor Fuel Company, controlled by Osgood, with ovens at Hastings and Gray Creek in the Trinidad district, and the American Smelting and Refining Company, which consumed the output of its own ovens at Durango.

The manufacture of coke centered in Las Animas County. In 1893, the Pueblo company's ovens were located at El Moro and Sopris, and they produced a smaller volume than did the plants at Crested Butte and Cardiff. Ten years later, six plants were operational in the Trinidad district. El Moro, Sopris, Starkville, Tabasco, Segundo, and Tercio accounted for four-fifths of the corporation's output, while the remainder was contributed by Crested Butte, Cardiff, and Redstone.[55]

Prior to 1900, the company maintained 250 ovens at El Morro, 222 at Sopris, and 140 at Starkville. That year, new stack models were constructed at Sopris and Starkville, raising the numbers to 272 and 190, respectively. Most of the output of these units was sent to the steel works, and about one-fourth was sold to smelters in Colorado and throughout the Southwest.[56]

Additional ovens were brought into production between 1901 and 1903 at Tabasco, Segundo, and Tercio. The largest of the new facilities was at Segundo, originally called Humoso, located approximately 17 miles west of Trinidad. In what was called the most complete plant west of Chicago, 800 beehive ovens turned out coke from coal mined at Primero, one mile away in straight-line distance, four by rail. The first battery, a double-row of 200 units, was fired in March 1902, and the last in January 1903, providing employment for 400 men, about one-sixth of whom were assigned to two washers. The product was shipped to the Minnequa Works and to smelters in the Rocky Mountain region.[57]

In Tercio Park, a plant consisting of 600 ovens that were arranged in two tiers of 300 units each, was completed in 1902. Provision was made for the construction of up to 1,200 more, but they were never built. Two washers processed slack and screenings from the Tercio and Cuarto mines. Most of the product (97 percent) was shipped to the steel works.[58]

At Tabasco, 302 beehive ovens were fired in 1901, using slack from the Berwind and Tabasco mines. The washer was not completed until the following year. More than nine-tenths of the output was sent to Pueblo.[59]

In the mountain district, plants at Crested Butte and Cardiff were the principal sources of coke for smelters in western Colorado. That industry's condition did not justify additional capacity at either place. However, the old Belgium ovens at Cardiff were replaced, commencing in 1900, by more efficient beehive models. The number of units were reduced from 250 to 165, but the volume of output was unchanged.[60]

Coke was also manufactured at Redstone in the valley of the Crystal River. Construction of 249 ovens began in 1899, and the first production was recorded the following year. Slack from Placita was consumed until fuel was available from Coal Basin. The decision to build the new plant was based not on a realistic appraisal of market conditions, but on John C. Osgood's hope of creating a model industrial community in the picturesque setting where he erected his private mansion. The ovens remained in operation less than ten years, victims of a declining smelting industry in western Colorado.[61] By the first decade of the twentieth century, the Minnequa Works was the region's largest consumer of coke, and it was not economical to haul that commodity from the mountain district to Pueblo.

The rapid expansion of mining and coking operations between 1899 and 1903 created tensions between labor and management that frequently erupted in strikes, most of which were of brief duration. In June of the former year, miners in Fremont County demonstrated for increases in wages. They were granted a 10 percent advance.[62] Similarly, miners and coke workers in Garfield County demanded and received higher rates of compensation.[63] Only at Crested Butte were concessions withheld. There the issue was recognition of the Western Federation of Miners. Rather than concede that point, the corporation closed the coking plant, throwing about 300 men out of work.[64]

Three hundred and fifty men employed at Sopris staged a walkout in May 1901, as a protest against the company's alleged failure to give them full weight for their coal, and against the high cost of the compulsory medical-care program. Following promises of a full investigation and corrective action, if warranted, the men returned to their jobs.[65]

The incident at Sopris was indicative of widespread discontent among miners in the southern field. Seven months later, that ill feeling culminated in a long and bitter strike that closed eight of the Colorado Fuel and Iron Company's mines. Although triggered by walkouts in the northern field and the Gallup district, the dispute was caused by local issues. The corporation charged that the walkout was organized solely for the objective of gaining recognition of the United Mine Workers of America. However, the men used the dispute to denounce the existing system of weighing coal, their inability to hire check weighmen, the use of scrip, restrictions on trade with independent merchants, the failure of wages to advance apace with the cost of living, and the "blacklisting" of union members.[66]

A settlement was negotiated on March 24, at which time the strikers began returning to their jobs. The Colorado Fuel and Iron Company made several concessions. The weighing system was modified to assure each man payment for a larger proportion of the coal that he mined. The men were permitted to hire check weighmen whose salaries were prorated among all workers and deducted from their monthly wages. And the cost of blasting powder was reduced by 10 percent.[67]

The reopening of the mines was delayed when the company refused to reemploy some of the strike leaders. After the miners threatened to renew the walkout, J. A. Kebler abolished the "blacklist," and peace was restored in the southern field.[68]

The strike of 1901 focused attention on the Colorado Fuel and Iron Company's hostility toward the United Mine Workers of America. In testimony before a joint-committee of the General Assembly, Osgood denounced that organization and blamed it for undermining the harmonious relations that had traditionally existed between the corporation and its employees. He and his associates, Osgood said, were always willing to work with local committess of miners, but they absolutely refused to "treat in any manner" with an autonomous national union.[69]

The failure of the southern operators, namely the Colorado Fuel and Iron and the Victor Fuel companies, to recognize the miners' union set the stage for another long and bitter labor dispute. On November 9, 1903, District 15 of the United Mine Workers of America, comprising Colorado, Wyoming, Utah, and New Mexico Territory, ordered a walkout in support of its demands for increased wages, an eight-hour day, a fair system of weights, biweekly paydays, and abolition of scrip. The real issue, however, was recognition of the union as bargaining agent for miners in the four-state region.[70]

Although peace was restored to the northern Colorado field within three weeks, strife continued in the southern counties until the middle of the following year. The union's hope for success was dispelled in March 1904, when Governor James Peabody ordered the Colorado National Guard to quell alleged acts of violence in Las Animas County. Militia units stationed at Engleville, Starkville, Sopris, Berwind-Tabasco, and Segundo enabled those mines to resume normal operations without fear of countermeasures by the strikers. The troops were finally withdrawn on June 11, by which time the walkout had for all practical purposes ended. Officially, the United Mine Workers of America did not terminate the strike until October 12, 1904.[71]

By the time the strike of 1903 began, John C. Osgood had lost control of the Colorado Fuel and Iron Company's management, and before it ended he had severed all connections with the enterprise which, as president, he had moulded into one of the West's largest industries. In little more than a decade under his guidance, the corporation's activities in the fuel trade had grown until it almost completely controlled the production of coal and coke in Colorado outside of the northern lignite field and in the major mining districts of New Mexico Territory.

By 1903, the company had in Colorado 23 mines which, together with 9 coking plants, employed about 6,000 men. These facilities were located in the southern field and mountain district, frequently in remote areas where, in the absence of towns, the company created on its own lands communities to which it attracted workers and their families. Because of geographical isolation and the nature of these mining camps, the residents were forced to depend on the employing corporation for services which, under the circumstances, it alone could furnish.

Osgood and his associates were sensitive to the needs of the employees. The management of the Colorado Fuel and Iron Company established elaborate medical, social, and education programs in the mill town and the mining camps, providing requisite services which workers, individually or through their communities, could not provide for themselves.

Altogether, these activities constituted an early form of welfare capitalism. This clearly anticipated the Colorado Industrial Representation Plan which the Pueblo company, at the urging of John D. Rockefeller, Jr., inaugurated in 1915 following still another disruptive coal miners' strike.

Trinidad Area Camps

Miner's dwelling at Morley.

Grade school at Morley. Five teachers are posing with students at the entrance to the school house, ca. 1910.

Miners' dwellings at Morley, viewed from an alley.

The Colorado Supply Co. store (left), camp doctor's office (center), and boarding house at Morley, ca. 1910.

The superintendent's residence at Morley.

Employees' club house at Morley, ca. 1915.

Morley Mine, originally called Katcina and then Swastika, was located 13 miles south of Trinidad and three miles north of the New Mexico border, on the mainline of the Atchison, Topeka and Santa Fe. The mine opened in 1907 to supply locomotives with fuel. When that market declined, coal was shipped to the by-product ovens at the steel plant. The east side drift was abandoned in 1929, and the west side drift on May 4, 1956, after all workable coal deposits had been exhausted. Pictured here are coal cars and miners outside the west side slope on the final working day.

Setting timbers to support the roof of the Morley Mine, May 4, 1956.

Mules performing underground haulage in the Morley Mine. Unlike most of the Colorado Fuel and Iron Corporation's mines, Morley was not highly mechanized because of the prevalence of gas in the working areas. Blasting, cutting and loading were done by hand, and mules pulled coal cars from working faces to collection points where they were removed from the mine by electric and rope haulage.

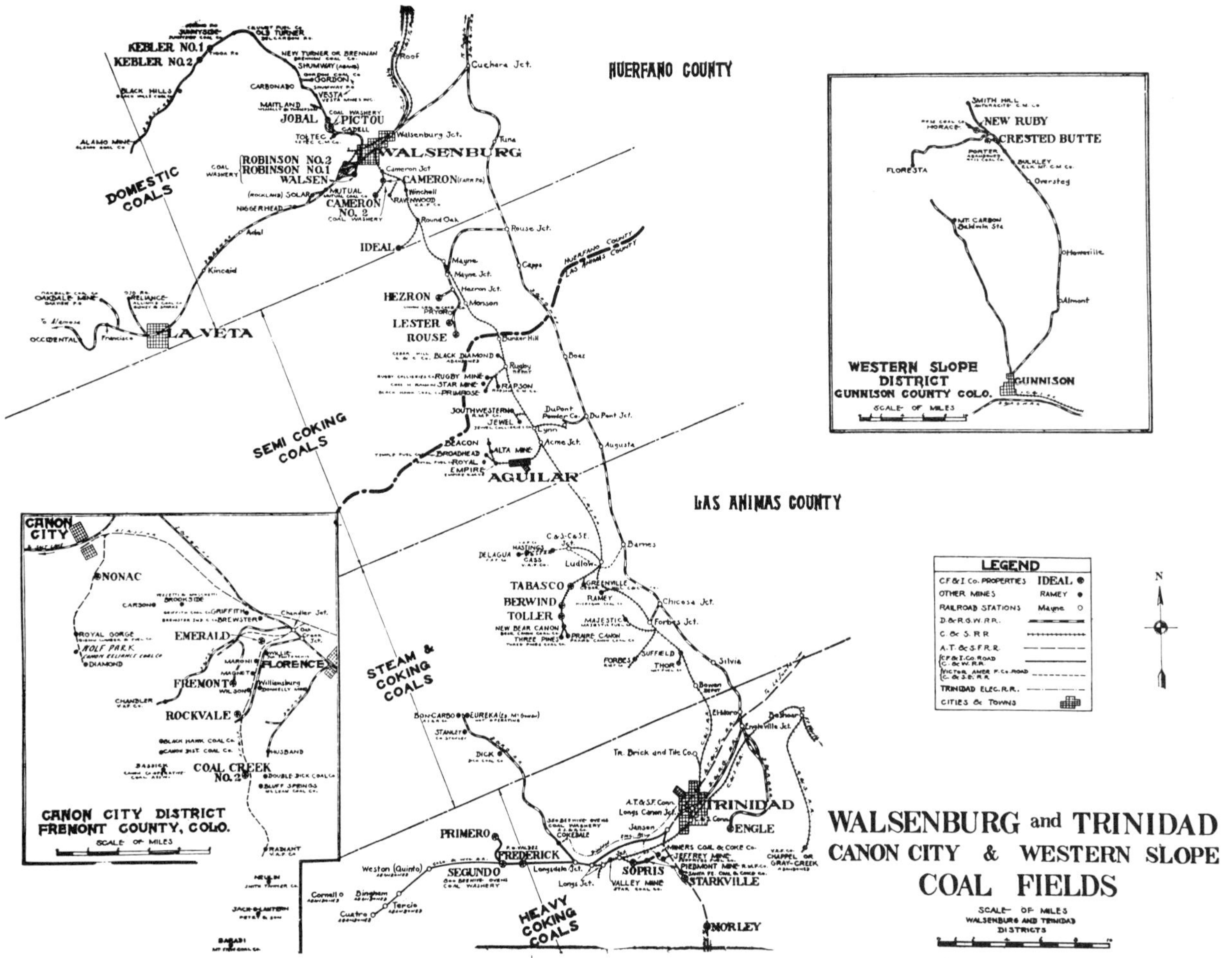

Walsenburg and Trinidad, Canon City, and Western Slope coal fields of Colorado.

Primero, founded in 1901, was designed by the Colorado Fuel and Iron Co.'s sociological department as a model community. Modern, comfortable houses, substantial school buildings staffed with highly qualified teachers, clubhouses, and other features enhanced the quality of life for coal miners and their families.

Primero Mine, located at the head of Smith Canon, 18 miles west of Trinidad. Known initially as Smith Canon Mine, then Purgatory, and finally, in 1902, Primero, the property was designed to be a large efficient producer of metallurgical coal. Most of the output was shipped to ovens at Segundo, 3.5 miles to the east. Production ended in 1925, by which time three drifts and a slope had yielded 8.2 million tons of coal.

Miners' wash house at the Primero Mine.

Colorado Supply Co. store at Primero.

Grade and high schools at Primero. The grade school was named for Alfred C. Cass, long-time vice-president and sale manager for the Colorado Fuel and Iron Co.

Camp dispensary at Primero.

Catholic Church at Primero. The town's population was in the early years composed largely of immigrants from Italy and the Austro-Hungarian Empire, men and women who were predominantly Catholic in their faith.

Superintendent's house, Primero. At the time this picture was taken in October, 1915, the residence was occupied by Joe Haske and his family.

A street in Primero, 1916.

Mine office at Primero.

Segundo, originally called Humoso, 14 miles west of Trinidad, was founded in 1901 when the CF&I Co. constructed the first of 800 ovens designed to convert coal mined at Primero into coke for the smelting industry and for the steel works. These beehive ovens continued in use after by-product ovens were completed at Pueblo in 1918. However, when the Segundo plant burned in 1929, it was not rebuilt. In this scene coal is being charged into ovens from the top, while coke has been drawn from doors onto the loading platform.

Frederick Mine, located at Valdez, 13 miles west of Trinidad, was opened by the Colorado Fuel and Iron Co. in 1907. Coking coal produced from a single drift was shipped mainly to the Segundo ovens. The mine was closed in 1960, and dismantled the following year. Total production was 20 million tons. Courtesy of The State Historical Society of Colorado.

Man trip cars at the portal of the Frederick Mine, 1950.

Mine office at Frederick.

Frederick store, 1915.

Tercio Mine and coke ovens, located 32 miles west of Trinidad, began operation in 1902. The output of the mine went principally to 600 ovens adjacent to the tipple. Mining ceased in 1915, by which time production exceeded 1.5 million tons. The ovens were used on a small scale after 1913, and shut down in 1918. Coke output amounted to 911,000 tons, almost all of which was consumed at the steel works.

10
Pioneer in Industrial Medicine

Southern Colorado was sparsely settled and only a short step removed from a frontier environment when the Colorado Coal and Iron Company was incorporated in 1880 for the purpose of erecting a steel plant and operating coal and iron mines throughout the region. The next year, that enterprise instituted a medical program to care for scattered settlements of employees. A hospital was established in Bessemer, and dispensaries were opened in each of the mining camps.

The Colorado Coal and Iron Company was a pioneer in industrial medicine. Its program, the largest in the state, was a model for other employers. In 1892, the corporation merged with the Colorado Fuel Company to form the Colorado Fuel and Iron Company — in its day, the industrial giant of the West. Ten years later, that firm owned or operated 40 properties in Wyoming, Utah, New Mexico Territory, and Colorado, including coal and iron mines, coking plants, railroads, a rolling mill in Laramie, and a fully integrated steel works in Pueblo. More than 15,000 employees and their families, totalling about 60,000 people, lived on or near the corporation's lands; they were dependent on the corporation for medical care and a host of social and educational services.

The man who by his genius and energy created and directed the Colorado Coal and Iron Company's medical plan, and that of its successor, was Dr. Richard Warren Corwin. A tall, handsome man who was contemplative by nature, he displayed a deep attachment to human worth and dignity. He dedicated himself to the service of others and to the corporate enterprises in whose behalf he devoted the whole of his professional career as a physician and surgeon.

Corwin came to Pueblo in the spring of 1881, to establish a hospital for the Colorado Coal and Iron Company's employees at the steel works and in mining camps throughout southern Colorado. The previous year, on completion of a two-year internship at St. Luke's Hospital in Chicago, he had planned to serve as physician-naturalist for a scientific expedition to South America. When that scheme failed to materialize, he applied for and received appointment as chief surgeon of the western enterprise. He retained that post until his death 48 years later.[1]

The only son of Walter Scott Corwin and Rhoda Little Corwin was born May 24, 1852, on a farm near Binghamton, New York. His youth was spent at Narrowsburg, on the Delaware River, and at Clifton Station, Virginia, near the battlefields of Bull Run and Cub Run. Those rural environments stimulated his interest in nature. At the age of 14 years, he was trained in taxidermy by John G. Bell, a noted authority who recognized the youngster's talent in that art.

Corwin completed his public school education at Port Jervis, New York, and enrolled at Cornell University in 1872. He supported himself by working as the school's taxidermist. He also participated in athletics as a member of the varsity rowing crew in 1874. That year, he withdrew from school for financial reasons.

Later the same year, Corwin was appointed curator of the University of Michigan's museum. He accepted the position because it enabled him to enroll in the school's medical courses. He was graduated in 1878, after which he completed his training in Chicago.

When Dr. Corwin reached his new home on April 16, 1881, he saw not one city but four. North Pueblo and Central Pueblo lay north of the Arkansas River; South Pueblo occupied the opposite bank, and Bessemer flanked the new steel works. The sight was unappealing. Streets were unpaved, sidewalks were largely nonexistent, the buildings that were mainly of adobe construction were in need of repairs, and saloons and gambling dens were common north of the river.

Heavy rains had turned the region into a swamp. The new steel works, then under construction, was nearly deserted; most of the men had been sent to the north end of Lake Minnequa, the company's reservoir, to repair a break in the dam. For the next two days, the medical officer served as foreman of a crew of laborers with wheelbarrows engaged in a futile effort to plug the hole with rocks, earth, and other materials.[2]

The Colorado Coal and Iron Company's hospital had been organized the previous year to provide medical care for the construction crews who erected a furnace, converter, rolling mill, and related facilities on the mesa south of the river. Originally the medical staff was located in building No. 8 in the residential section that fronted on the steel works. Early the following year, No. 7 was taken over in order to cope with the expanding case load.[3]

In 1882, the temporary hospital along "Furnace Row" proved inadequate for the company's needs. That spring, a threatened outbreak of smallpox prompted emergency measures. A pest house was constructed south of Bessemer to isolate victims, and construction workers and other citizens were vaccinated.[4] Those tactics prevented the spread of the disease. Later the same year, the medical staff was nearly overwhelmed by an epidemic of "mountain fever." A mild form of typhoid fever, the sickness reached serious proportions with a total of 133 cases. That experience convinced general manager Appleton H. Danforth that the company needed a larger hospital.[5]

As early as April 1881, Danforth had discussed the prospect of a joint hospital with David C. Dodge, vice-president and general manager of the Denver and Rio Grande Railway Company. They agreed that their respective enterprises, and perhaps others, should form an association to erect a hospital in South Pueblo. It was to be a two-story brick building with a central section for offices and a 24-bed wing designed so that other units could be added when needed. Danforth suggested that Dr. Corwin should be placed in charge, and that he should be entrusted with the task of planning with the architects the design of the facility.[6]

Plans prepared by Holabird and Simonds, of Chicago, prompted a brief but heated debate over the proper facility required by the two companies. Using Dr. Corwin's suggestions, the architects designed a building with only "absolutely necessary" accomodations which were to cost, with furnishings, about $9,000. Dodge approved, but later reversed himself when Dr. F. J. Bancroft, chief surgeon for the railroad, insisted upon the adoption of the so-called "Pavillion Plan" in which separate wards were located around and detached from the main hospital unit. After a series of exchanges setting forth arguments for and against that scheme, Dodge reaffirmed his original endorsement.[7]

The new hospital was erected on East Abriendo Avenue. Reinerston and Person, contractors, began in October 1881 the construction of a building 100 by 30 feet in dimensions. Inclement weather delayed and eventually halted the work until April 1882. The structure was completed and furnished before the close of that year.[8]

The 30-bed complex opened in December 1882 with a small staff. Dr. Corwin was chief surgeon, assisted by Dr. J. A. Pascoe, a recent arrival from the East. Will H. Taylor, a druggist, was in charge of the dispensary. At this time, female nurses were introduced. The previous year C. K. Kastholm and George Wise had come to Bessemer as male nurses.[9] Misses F. Richardson and A. Couton were the first of a number of women who shared the nursing function until 1897, when men were excluded from that duty.

The two companies shared the cost of constructing and furnishing the hospital, and the coal and iron enterprise donated the tract on which the building was located. Thereafter, as set forth in the original agreement, each corporation owned an undivided one-half interest in the structure, contents, and grounds. Each had the right to send employees to the hospital for medical care, and operating costs were shared in proportion to the number of people admitted from each firm.[10]

The cost of the building and furnishings exceeded the original estimates. The architects had been instructed to submit plans for a structure to cost, with contents, no more than $9,000. The actual cost was $19,076.44. The Colorado Coal and Iron Company, using its own laborers, erected the building. Danforth intimated that this may have been one reason why the cost was greater than anticipated. Dodge, after examining an itemized statement for construction and furnishings, approved payment of the railroad's share. Because of a change in administration at that time, prompted by General Palmer's resignation from the presidency, a voucher was not processed. Although a settlement was not effected for many years, the two companies jointly used the facilities.

The medical program was fully implemented in 1882, with the placement of physicians in the company's mining camps. Dr. A. T. King resided in Coal Creek, Dr. John Grass in Trinidad, Dr. Martin in Walsenburg, and Dr. Snyder in Crested Butte. From Salida, Dr. J. W. O'Connor, the railway's surgeon, made calls at the iron mines near that community and Villa Grove. Major Cochran, who was stationed at Fort Garland, attended the sick and injured at the Placer iron mine.[11]

A. T. King, a graduate of Northwestern University and the internship program at St. Luke's Hospital in Chicago, remained at Coal Creek until 1884, when he resigned to enter private practice in Pueblo. Subsequently, he maintained close personal and professional ties with Dr. Corwin, serving for many years as a consultant at the company's hospital in South Pueblo. In 1895, King was elected mayor of Pueblo as a reform candidate on the Citizen Party ticket, and served a single two-year term.[12]

The Colorado Coal and Iron Company assessed each of its employees a monthly fee of one dollar to pay its share of the hospital's operating expenses and to maintain dispensaries in the mining camps. Initially, that policy was resisted by some coal miners who believed that the deduction was excessive, and that the care they received from resident physicians was inferior to what they could obtain from private doctors.

The medical fee was a major issue in an early labor dispute. During the prolonged strike of 1884, some miners stated that the company had derived large profits from the hospital and camp dispensaries.[13] Reacting to that charge, Danforth at first agreed to place the administration of the hospital fund under the supervision of a board of managers composed of one representative from each mine and an equal number of the corporation's officers. Later he abandoned that proposal and in its place offered employees the option of accepting or rejecting the program. If the men chose the latter option there were to be no further reductions from their wages; nor were they to have the privilege, without additional expense when sick or injured, of seeing the local physician or of entering the hospital in South Pueblo.[14]

On the other hand, if they elected to retain the program, certain reforms were to be instituted. For example, monthly and quarterly reports were to be posted at all of the mines. If the fund accumulated a surplus, the assessment was to be reduced, or the money used to create a fund from which the families of men killed in the underground workings were to receive benefits.

When the referendum was presented to employees, the response was overwhelmingly in favor of retaining the existing medical program. At Walsenburg, the vote was 156 in favor and only 1 opposed.[15] At El Moro, only 2 of 126 men voiced objections to the current plan.[16]

In subsequent years, as the Colorado Coal and Iron Company and its successor expanded operations at the Bessemer Works or in the mining camps, the medical program was enlarged to accommodate additional employees. In 1890, the capacity of the hospital was increased to 45 beds. Five years later, an operating room and dispensary were added. A large wing constructed in 1897 provided an additional 35 beds. Remodeling projects in the next five years raised the total to 112. At that point, further expansion was out of the question because there was no more land at the East Abriendo site.[17]

A dispensary at the steel works, although under Dr. Corwin's supervision, was distinct from the hospital. While the function of the latter was to cure diseases and treat major injuries, the former served mainly to prevent sickness and to administer promptly to ailments which, if neglected, might prove serious. Physicians and attendants were on duty 24 hours a day. For the most part, they treated mild afflictions such as colds and coughs and small cuts and bruises resulting from minor accidents. Emergency care was provided for victims of accidents prior to sending them to the hospital.[18]

On January 1, 1901, the dispensary was moved from a site near the pipe foundry within the plant to a location adjacent to the company's new offices on Canal Street outside the west gate. A six-room Spanish mission style building contained waiting, drug, consultation, surgical, and storage rooms, in addition to sleeping quarters for the resident physician.[19]

In 1902, when the employees at the Minnequa Works exceeded 5,000 in number, the dispensary handled an average of 75 cases daily, or 23,000 in the year. The staff, under the direction of Dr. C. V. Marmaduke, who had been appointed to that position in 1896, included Doctors O. F. Adams and A. W. Scarlett. In addition to their duties at the mill, they made house calls to attend to the sick or injured whose conditions did not warrant admission to the hospital.[20] Scarlett was also the resident physician at Lime, a quarry town six miles south of Pueblo.

Similar facilities existed in each of the mining camps. Each unit was staffed by a resident physician whose salary and expenses were defrayed from the hospital fund. The "camp doctor" was required by the company to treat all illnesses except venereal diseases and all injuries other than those that resulted from violations of state laws and municipal ordinances. Patients who required extensive or special care were sent to the hospital, to which no one was admitted except on a doctor's order.[21]

Twenty-nine physicians resided in the mining camps in the summer of 1902. They were the following: Albert L. Trout, Berwind; R. E. Holmes, Brookside; Leman Gibbs Clark, Cardiff; E. D. Harper, Catalpa, New Mexico; W. E. Ashby, Coalbasin; Arthur A. Eddy, Coal Creek; J. W. Rockefeller, Crested Butte and Floresta; T. J. Forhan, who was also the company's assistant chief surgeon, El Moro and Engleville; Charles F. Beeson, Fierro, New Mexico; James Jay Pattee, Gibson, New Mexico; T. M. Ahlquist, Hezron; A. B. Hamilton, Laramie, Wyoming; A. W. Scarlett, Lime; S. C. Clarke, Madrid, New Mexico; Orland P. Shippey, Villa Grove; T. D. Baird, Pictou; William M. Ogle, Primero; Angus Taylor, Redstone and Placita; W. A. Williamson, Rockvale and Fremont; Walter S. Chapman, Rouse; D. Guy Leach, Segundo; H. C. Dyer, Spring Gulch and Sunlight; C. O. McClure, Starkville; James Edward Lowery, Sopris; Harry C. Lee, Sunrise, Wyoming; O. W. Cosby, Tabasco; W. L. Conway, Tercio; J. H. Davis (Colorado and Wyoming Railway), Trinidad; and D. W. Matthews, Walsen and Robinson.[22]

After the hospital adopted a policy in 1897 of employing exclusively women in nursing roles, Jennie C. Sutcliffe, a member of the staff of St. Luke's Hospital in Chicago, was hired to establish a two-year training school for young ladies. The Colorado Fuel and Iron Company School of Nursing was formally incorporated in May 1899, by Doctors Corwin and W.T.H. Baker and Miss Sutcliffe. Classes had commenced the previous fall, and the first two graduates received certificates in 1900.[23] Thereafter, the enrollment expanded at a rapid rate.

Under the direction of Miss Jennie S. Cottle, the school adopted a three-year curriculum in 1901. The new chief nurse was a graduate of Wellesley College and had received her professional training at Harper Hospital in Detroit. She was assisted by a staff of four registered nurses.[24]

The school year commenced September 1 and terminated June 24. Only women of superior education were admitted. Each student received board, lodging, and an allowance of $8.00 per month for uniforms and textbooks. In the first year, pupils served as junior nurses, the second year as intermediate nurses, and the final year as senior and head nurses. During the last year, they were compensated for duties performed in the hospital.[25]

By the close of the century, it was apparent that the medical facilities had to be expanded. The acquisition of additional coal mines and the opening of new iron deposits added large numbers of men to the company's payroll. The new employees threatened to overwhelm the hospital. In order to cope with the increased load, so-called "retaining hospitals" were created at Fierro, Gallup, and at Sunrise. Those units, maintained under the supervision of the resident physicians in the camps, were capable of treating a dozen or more patients at a time. The seriously sick or injured were forwarded to Pueblo.[26]

As early as 1899, following the decision to double the output of the steel works, company officials had recognized that the central hospital had to be enlarged again. The next year, when tents had to be erected to accomodate a record number of patients, Corwin pressed for immediate action.

Although there was no room for additional buildings on the grounds of East Abriendo Avenue, it was possible to erect new facilities across the roadway and to connect the two units with a tunnel. In discussing this plan with company officials, Dr. Corwin was asked by Julian A. Kebler if that was the best solution to the problem. Without hesitation, the chief surgeon replied that it was not and that a new location near Lake Minnequa was far more desirable. The two men immediately drove to the proposed site.

While examining an empty tract located east of the lake on Minnequa Avenue, Kebler agreed that it was an ideal spot for a modern medical facility suited to the company's enlarged payroll. Turning to his friend, he said with a grin, "You go and study hospitals and when you return, build a hospital, and if it does not suit we will fire you."[27] That was the authorization for a new complex of buildings. Corwin was determined to make them modern and functional in every respect.

After studying plans of hospitals in the United States and abroad, Dr. Corwin returned to Pueblo to work with Jacob M. Gile, architect for the firm of Sterner and Williams, in designing the new buildings. A contract was awarded in November 1900, construction began in April 1901, and the complex was occupied in August 1902, at which time patients were transferred from East Abriendo Avenue.[28]

On August 6, one day prior to the official opening, Minnequa Hospital was placed on public display. Nearly 4,000 people examined the buildings and grounds. Probably the vast majority agreed with Dr. Corwin's assessment of the new facility. The employees of the Colorado Fuel and Iron Company, he said, have the best hospital in the world, not because it is the largest or the most ornate, but because it has an expert staff plus all the equipment and features required to treat the sick and injured.[29]

Occupying a 20-acre tract on Minnequa Avenue between Lake and Orman Avenues, the hospital was erected at a cost of about $300,000. The complex included 13 Spanish mission style buildings, constructed of brick, covered with cement, decorated with iron balconies on the second floors, and roofed with red tiles. Completed and ready for occupancy were the executive and housing units, three wards, surgical or operating unit, ward for communicable diseases, physicians' residence, recreation hall, stable, well house, and the light and power plant. Four buildings remained to be constructed: the nurses' home, pathological and laboratory unit, convalescents' lodge, and chapel.[30]

The hospital was designed with the principal structures as a central cluster connected by corridors. The main entrance, on Minnequa Avenue, was to the executive building, which contained administrative offices. Directly to the rear was the House Department with dining rooms, maintenance services, and storage areas. To either side of those units were wards for patients: Nos. 1 and 3 to the west (left) and No. 2 and the surgical building to the east (right).[31]

The wards were two-story structures with single- and multiple-patient rooms, all of which were above the ground with windows that faced to the north and south on landscaped grounds. There were no courtyards or "inside rooms."

Each floor in the wards contained six single- and six four-bed rooms, four bathrooms, two lavatories, a maid's closet, diet kitchen, and nurses' room. Single units were for the seriously ill patients. By limiting occupancy in the larger units, Dr. Corwin hoped to segregate people according to the nature of their afflictions, ethnic origins, or dietary needs. Throughout each structure, showers were employed rather than bathtubs, what the chief surgeon called "articles of a barbaric age," and all plumbing was removed from the walls for easy cleaning.[32] Brick partitions separated rooms in order to retard the flow of noise.

Enclosed corridors that connected the main buildings served as sun parlors. Since they were laid out in a north-south direction, they provided a maximum exposure to solar light.

The surgical building contained supporting facilities and an operating room that was completely lined with lead. That material had been employed experimentally in a similar role in the "old" hospital for seven years with satisfactory results. Dr. Corwin viewed it as the ideal lining for a sterile room. The metal in the floor weighed 12 pounds per square foot, twice as much as that in the walls and ceiling. Light was admitted only from the north, an arrangement designed to prevent "cross lights" which frequently annoyed surgeons during operations.[33]

One of the unique features of the hospital was the almost complete absence of stairs. A driveway ran from Minnequa Avenue to the main entrance which was halfway between the first and second stories of the executive building. This made all floors in the wards readily accessible by means of inclines that ascended or descended at a moderate rate of one foot in seven. On admission, a sick or injured person could be placed on a wheeled stretcher and conveyed to any room in the three wards without resort to stairs or elevators. Ambulatory patients found that the inclines made it easier to move from one floor to another.

When the plans for Minnequa Hospital had been drawn, the architect wanted to use stairs rather than inclines on the grounds that the former were more economical. On hearing of this, Dr. Corwin retorted, "If that is the only objection, why not use ladders instead of stairs!" The chief surgeon had his way. Inclines were incorporated into the final plans for the buildings.[34]

The medical staff at Minnequa Hospital, in addition to the chief surgeon, was composed of nine physicians and surgeons. Dr. William T. H. Baker, a native of Illinois and a graduate of Northwestern University Medical School in 1896, had been appointed to the Bessemer dispensary that year, then promoted in 1897 to the post of superintendent of the hospital.[35]

Three specialists served on the medical team. Dr. Edgar M. Marbourg, a native of Philadelphia and a graduate of that city's Jefferson Medical College, treated ailments of the eyes, ears, nose, and throat. Dr. William H. Hoch, a product of the University of Pennsylvania, was an expert on the diseases of the ears, nose, and throat. Dr. Herbert Smith Olney, a graduate of the University of Michigan Medical School, completed his internship at the Colorado Fuel and Iron Company Hospital, at the expiration of which, in June 1902, he was appointed the staff pathologist.[36]

Five interns completed the medical roster. Dr. Ortus Fuller Adams was a graduate of the Medical Department of the State University of Iowa; he served for a time in the Bessemer dispensary. Doctors Howard Turner Carriel, Jay Thomas Downing, and Charles E. Smith were products of Northwestern University Medical School. Dr. Lee Masten Francis was a graduate of Chicago's Rush Medical College.[37] Dr. William Senger, a native of Port Jervis, New York, and a graduate of the Yale University Medical School, was destined many years later to succeed Corwin as the Colorado Fuel and Iron Company's chief surgeon.

Local medical men were used as consultants. Dr. Pembrose R. Thombs, who held that position for more than two decades prior to his death in June 1902, had been superintendent of the Colorado Insane Asylum from 1879 to 1899. Dr. Hubert Work, founder of Mount Pleasant Hospital (later Woodcroft) in Pueblo, became one of Corwin's part-time experts in 1898. Dr. A. T. King, a local physician, had occupied a similar post since 1884. Dr. W. E. Varley, a Pueblo dentist, was appointed to the staff in 1890.[38]

Dr. Corwin, a surgeon of extraordinary talent, was widely recognized and acclaimed by colleagues throughout the medical profession. In addition to his duties as director of the Colorado Fuel and Iron Company's hospital and dispensaries, he held, prior to 1901, appointments with other companies and institutions. He was at one time or another surgeon for the Denver and Rio Grande, the Denver and New Orleans, and the Colorado and Southern railroads, the Colorado Smelter of Pueblo, and the Colorado Insane Asylum. He also held the rank of professor of surgery on the faculty of the University of Colorado Medical School, and for five years served as a member of the Colorado Board of Health.[39]

In addition, Dr. Corwin was for many years an active member of the Colorado National Guard. Entering the service in 1884 as a first lieutenant, he rose to the rank of colonel and served as surgeon general before resigning in 1898. He served again in the latter capacity from 1903 to 1904, and from 1907 to 1908.[40]

During his lifetime, Dr. Corwin received from his peers in the medical profession numerous honors and distinctions. He served as president of the Pueblo County Medical Society, the Colorado Medical Society, and the American Association of Railway Surgeons. He was at one time vice-president of the Colorado Branch of the International Congress on Tuberculosis and a director of the National Association for the Study of the Prevention of Tuberculosis. He was also a charter member of the American College of Surgeons and remained active for a long time in the American Medical Association, the Association of Military Surgeons, and the American Association of Industrial Physicians and Surgeons.[41]

An abiding interest in the betterment of human life prompted Dr. Corwin to play an active role in community affairs. He was a supporter of literary endeavors and was instrumental in organizing the McClellan Public Library in Pueblo, serving as the first executive officer of the library association that administered that institution. For more than 40 years, he was a member of the Board of Education, District 20 (Pueblo south of the river), a record for sustained service that was unmatched in that day. For eight years, he served on the Colorado State Normal School Board; for several years, he was a member of the Colorado State Board of Agriculture. A long-time member of the Colorado State Conference of Charities and Corrections and of the Colorado State Prison Association, he served both groups for a time in the capacity of president.

On June 19, 1929, death closed Dr. Corwin's long association with the Colorado Fuel and Iron Company. The following day, the corporation's board of directors changed the name of the firm's major medical facility in Pueblo to Corwin Hospital in order "to link forever to this splendid institution the name of the man who from 1881 to 1929 gave himself unreservedly to its development."[42]

Industrial Medicine

The Colorado Fuel and Iron Co.'s Minnequa Hospital shortly after it was opened in 1902. Courtesy of The State Historical Society of Colorado.

Main gate to Minnequa Hospital grounds, Pueblo.

Convalescent and nurse on Minnequa Hospital grounds. Front entrance to the main hospital building is in the background.

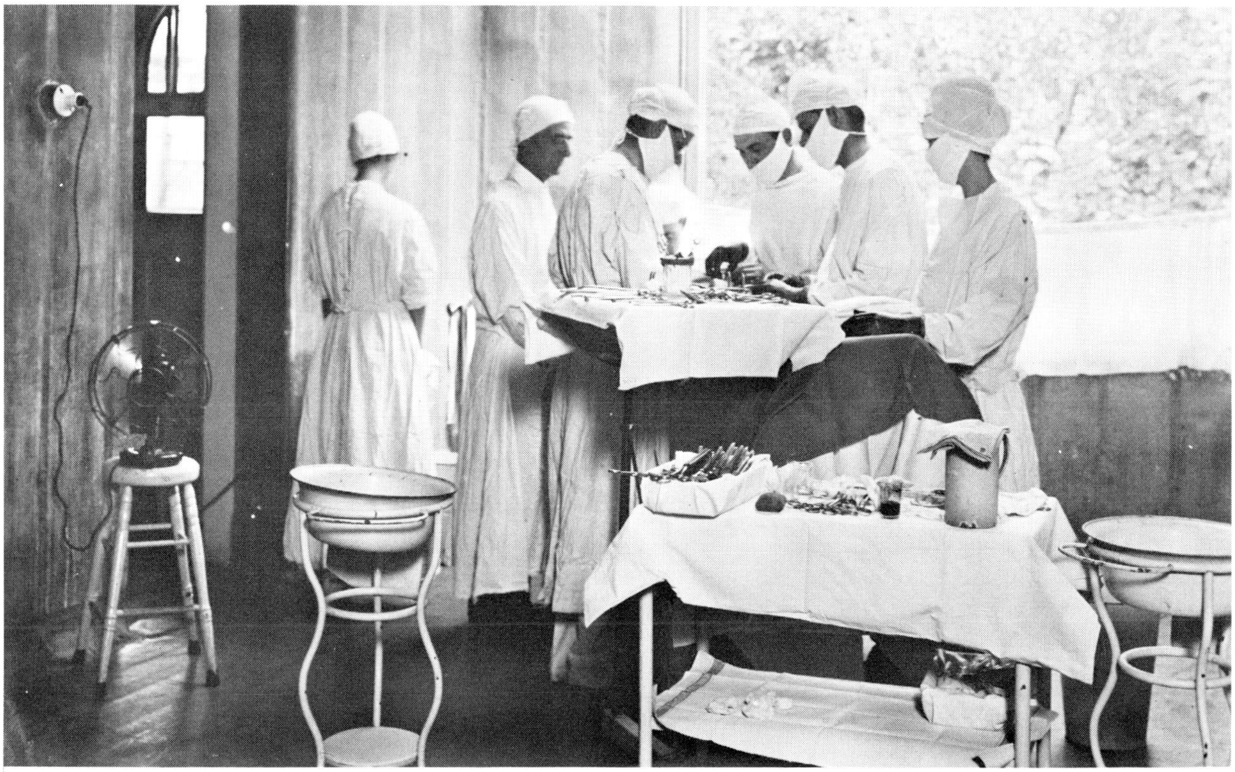

Operating room at Minnequa Hospital. The floor, walls, and ceiling are lined with lead.

Convalescent on grounds of Minnequa Hospital, 1916.

Convalescents on Minnequa Hospital grounds.

Rustic bridge on grounds of Minnequa Hospital, Pueblo.

Dr. Richard Warren Corwin, chief surgeon of the Colorado Coal and Iron and the Colorado Fuel and Iron companies, 1881-1929. Corwin was an outstanding pioneer in the development of industrial medicine and sociology.

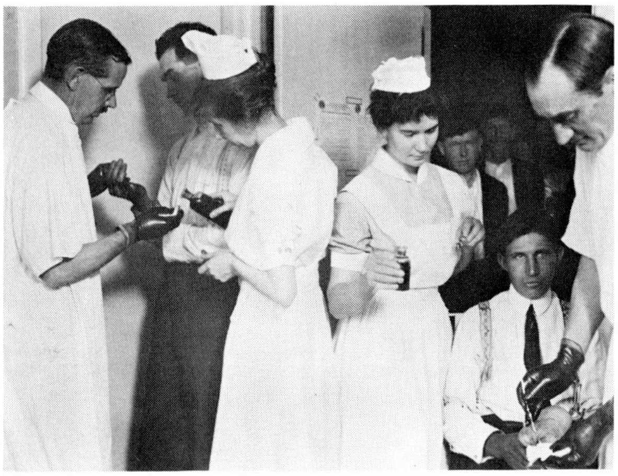

Treating patients at Minnequa Hospital.

Steel works dispensary, located east of the main administration building on Canal Street.

11
Industrial Sociology

Minnequa Hospital was only a partial fulfillment of the dream that had been born in 1881, when Dr. Richard W. Corwin opened a two-room hospital on Furnace Row in the new town of Bessemer. Twenty years later he expanded upon his medical career by inaugurating new programs designed to improve the quality of life and the environment in the Colorado Fuel and Iron Company's mill towns and mining camps.

Julian A. Kebler announced July 25, 1901, the creation of the Sociological Department. Dr. Corwin, in addition to his duties as chief surgeon, was given "general charge of all matters pertaining to education and sanitary conditions," as well as any programs designed to improve the conditions of employees and their families.[1] The objective, according to the general manger, was not paternalism but to put workingmen "in the way of obtaining information that will arouse their ambition and make them desirous of doing the best they can for themselves, as well as for their employer."[2]

The timing of Kebler's announcement suggests that the decision to organize the Sociological Department may have been prompted in part by the coal miners' strike of 1901. The General Assembly's investigation of conditions in the mining industry had been especially critical of the Colorado Fuel and Iron Company's labor policies.[3] The new activity may have been inaugurated in an attempt to improve the corporation's public image while at the same time reestablishing harmonious relations with employees.

Prior to 1901, the fuel and iron company's management had generally ignored conditions in the coal towns. The welfare of miners and their families, if of concern to the corporation at all, was delegated to local superintendents who, because of their positions, were preoccupied with problems of producing and shipping coal and coke. Consequently, the towns were for the most part neglected and unattractive. Dwellings were often makeshift contrivances, and community services were often nonexistent.

In establishing the Sociological Department, the company recognized that its varied activities were supported by a work force which, because of its size, location, and composition, required special programs. In the mining camps, many of which were isolated from other population centers, the corporation was invariably the only employer and often the only landowner. It was, for that reason, the dominant influence that shaped the lives, habits, and attitudes of all who lived in the communities. This fact inspired Dr. Corwin, with the support of management, to make the enterprise a social as well as an economic force in the lives of employees and their dependents.

In the decade prior to 1903, the number of men on the company's payroll had increased from about 5,000 to more than 15,000. Approximately two-fifths of that number were employed in the steel works; the balance were in production facilities scattered throughout a four-state region.[4] In Colorado alone, there were 29 settlements, of which twenty-three were supported by coal mines, one by an iron mine, and three by coking plants, and two by limestone quarries. Additional mines were located in Wyoming, Utah, and New Mexico Territory.

About one-half of the Colorado mining camps had come into existence in the previous ten years. In the Purgatoire Valley, these included Tabasco, Primero, Segundo, Tercio, and Cuarto. Santa Clara (New Rouse) and Hezron were near Walsenburg. Calcite was in Fremont County. Floresta was located west of Crested Butte; and Placita, Coalbasin, and Redstone were in Pitkin County.

The communities had to provide a host of services for residents. The responsibility for housing, police and fire protection, sanitation, public health, education, and recreation ultimately fell on the employer, because it was in almost every instance the sole owner of real estate within the towns.

The expansion of older mines and coking plants created similar problems in established communities. As new workers were hired, populations increased, creating the need for enlarged services which, because of its dominant position, the company had to provide, or at least to take the initiative in making them available. The heterogeneous character of the residents made it difficult if not impossible for them to work together in organizing community functions.

Prior to 1890, the overwhelming majority of the people in the camps were English, Welsh, and Scots, or native American descendants of those stocks. While each group possessed unique characteristics, their cultures were similar. All spoke English, adhered to Protestant faiths, and shared common aspirations.

In the final decade of the nineteenth century, the composition of the camps changed. This had been underway for many years. The populations of southern towns had been diluted as early as 1879, when miners of Mexican ancestry were hired at Engleville, and subsequently at El Moro and in the mines near Walsenburg.[5] During the labor disputes of 1884 and 1885, Negroes, Italians, and "Austrians" appeared for the first time in large numbers. The latter were people of Slavic origins from the Austro-Hungarian Empire.

By the opening of the new century, recent immigrants were clearly in the ascendency. In Fremont, Huerfano, and Las Animas counties, they comprised two-thirds or more of the coal diggers, loaders, and general laborers in the company's mines and coking plants. The principal elements were from southern and eastern Europe. The former were Greeks, northern and southern Italians, and Spanish. Those from eastern Europe included Bohemians and Moravians, Bosnians, Bulgarians, Croatians, Dalmatians, Herzegovinians, Lithuanians, Magyars, Montenegrins, Poles, Rumanians, Russians, Ruthenians, Servians, Slovaks, and Slovenes.[6]

Because of the new elements, most mines had diverse working forces. At Engleville, the oldest of the camps, the payroll in 1902 consisted of 140 Italians, 94 Mexicans, 64 Slavs, 12 native Americans, and 60 British, the latter made up of English, Scots and Welsh.[7] Farther north, the employees of the Walsen and Robinson mines were equally mixed, including 60 Italians, 47 native Americans, 45 Negroes, 32 Japanese, 32 Scots, 23 Mexicans, 21 Slavs, 8 Germans, 8 English, 7 Irish, 4 Welsh, 2 French, and 1 Swede.[8]

Some employee-oriented programs had been carried on by the Colorado Fuel and Iron Company and its predecessors for many years prior to 1901. Special rooms had been set aside at Bessemer and in the mining camps for reading and recreation.[9] Musical groups, mostly brass bands, had appeared in Crested Butte and Bessemer in the decade of the eighties.[10] In 1892, kindergartens were established, the first at Sopris, and others in nearby towns.[11]

Not until 1901 was the concept of industrial sociology accepted by the company and applied throughout its properties. To that end, Dr. Corwin organized a dedicated and experienced staff. As assistant superintendent, H. J. Wilson, and later Walter Morritt, assumed charge of the day-to-day conduct of all programs. Mrs. Margaret S. Grabill, of Trinidad, was supervisor of kindergarten and domestic science courses. Lawrence Lewis served as editor of a weekly (later monthly) publication. C. C. Everhart was responsible for construction of facilities. In addition, there were eight instructors for kindergartens, two for domestic sciences, eight for adult classes, six managers of reading rooms and club houses, and directors of the Minnequa band and the recreation hall at Starkville.[12]

In the summer of 1901, the Sociological Department launched programs in five specific fields: education, social training and industrial training, housing, and communications. These endeavors, according to Corwin, were not a passing fancy or an expression of sentiment, but implementation of a functional science called sociology. While in the interest of the company, its objective was to benefit employees and their families by educating children, improving relations in the home, moulding better citizens, and fostering an appreciation for labor. In some instances, evidence of improvement could be expected immediately; however, since habits were not easily altered, the full impact could be accurately measured only over a period of years.[13]

Educational programs were instituted for children, mothers and housewives, working adults, and the elderly. In the first category, stress was placed on kindergartens. Dr. Corwin believed that although it is difficult to change the manners and values of mature people, youngsters are tractable, easily managed and moulded, and without habits that needed to be recast or corrected. He felt that instruction for children between the ages of three and six years, the most impressionable period of life, was ultimately the key to social betterment.[14]

The Sociological Department's kindergartens sought to develop strong, refined characters in children and also the momentum to carry them over obstacles certain to be placed in their way by the environment in which they lived. The classes, operated initially at the company's expense, placed each child and nationality on an equal standing and, while recognizing individual differences, attempted to inculcate sympathy, unselfishness, and an appreciation for equal rights, the essentials of a democratic society.[15]

Within three years, kindergartens had been established at Sopris, Rouse, Starkville, Engleville, El Moro, Pictou, Walsen, Crested Butte, Redstone, Berwind-Tabasco, Primero, Segundo, and Tercio. During the 1903-1904 school year, school districts in the last four localities assumed the costs of the classes.[16]

In conjunction with the kindergartens, the houses provided for teachers served not only as residences during the school year but also as headquarters for social work in the camps. The furnishings, all thoroughly practical and sanitary, were deliberately chosen so as to serve as examples from which local housewives learned at first hand what could be accomplished at modest costs in their own homes. The model abodes were maintained in Berwind-Tabasco, Primero, Segundo, Engleville, Rouse, Pictou, and Crested Butte.[17]

Although the education of older children was left to local school districts, the Sociological Department offered many forms of assistance. Corwin believed that the Colorado Fuel and Iron Company, as the largest taxpayer in the camps, was entitled to make reasonable demands with respect to buildings, the competency of teachers, and free textbooks for students. At his insistence, the State Superintendent of Public Instruction supplied uniform courses of study which were adopted throughout the mining camps. Because of this system, children encountered few problems of adjustment to the academic program if their parents moved from one to another community.[18]

On request, the Sociological Department assisted local school districts by soliciting teachers to fill vacancies. It also advanced monies, on occasion, to pay salaries and to finance construction of new buildings. At Madrid, New Mexico Territory, in 1903-1904, the instructor for the primary grades was paid by the department.

The Sociological Department prepared a standard plan for a plain but serviceable school house which was widely used in the coal towns. Although the size and ornamentation differed from one locality to the next, the structures were usually of similar design.[19] A spacious vestibule was utilized as a coat room and contained the staircase to the upper level. Swinging doors led from the entry to a classroom with seats for about 50 pupils. If two rooms were located on one floor, folding partitions rather than walls were frequently installed. In this way, the units could be converted into a large hall when that was needed.

Buildings of this type were frequently named for officers of the company. Osgood School was located in Rouse. The structures at Pictou, Primero, Segundo, Tercio, and Berwind were named, in order, Kebler, Cass, Jerome, Beaman, and Corwin.[20]

In many mining camps and in Bessemer, the Sociological Department regularly conducted courses in domestic sciences. Classes in the preparation of foods, sewing, and housekeeping were usually well attended. Many daughters of non-English-speaking workers sought instruction in these fields; their mothers generally did not, probably because of the language barrier.[21]

Night courses afforded opportunities for instruction in English, with attention to reading and writing as well as the art of speaking. On occasion, the curriculum included geography and history. This part of the Sociological Department's work was self-supporting. All students were charged $1.00 per month to compensate the teachers and to defray the cost of renting or maintaining rooms.[22]

The stress on night classes, whether in Bessemer or the camps, was a matter of practical necessity. In 1901, there were among the employees 32 nationalities; 27 different languages were spoken, excluding dialects. Only a small percentage of the recent immigrants spoke any English. Even fewer could write. At one mine, which may not have been an exception, nearly every man made a mark rather than a signature in receipt of his monthly wage.[23]

Educational programs were expanded in 1903, when the old hospital on East Abriendo Avenue was renovated and reopened as the Pueblo Normal and Industrial School; it operated under the auspices of the Sociological Department. During the summer months of each year, teachers from the coal camps enrolled in academic and special training classes. In addition to the regular curriculum, instruction was offered in cooking, sewing, music, art, dancing, basketry, weaving, and other skills required for conducting social, industrial, and recreational programs in the mining towns. On occasion, prominent educators lectured to the students. The Minnequa Symphony Orchestra and the Minnequa Steel Works Band presented concerts.[24]

The Colorado Polytechnic Association, formed in July 1903, converted a portion of the old hospital building into a lounge, with reading and lecture rooms. Composed of engineers, superintendents of departments, and managers of plants within the steel works, that organization sponsored social and educational activities, with emphasis in the latter instance on advances in the technology of steel making.[25]

The old structure on East Abriendo Avenue also contained an Industrial Home that offered jobs to men who had been permanently injured in the mills or mines and to widows and children of deceased employees. By manufacturing products for which there was a demand in the local market, they hoped to gain enough income to escape dependence on charity.[26] Mattresses, brooms, hammocks, furniture, rugs, carpets, and lace were sold in Pueblo and in the mining communities.

Industrial activities were also instituted in many of the mining camps. At Engleville and El Moro, lace was produced in quantity. It was readily marketable and brought a cash return commensurate with the labor expended in its manufacture. Foreign-born housewives participated in the program as a means of augmenting their husbands' incomes. The women of Primero developed basket weaving as a household industry.[27]

Training in practical subjects was an important part of the Sociological Department's work with young people in the mining camps. Since most of the children were destined to become miners, or the wives of coal diggers, Dr. Corwin insisted that they be exposed to the skills requisite for their roles in life. Girls were taught to maintain neat homes and to prepare savory meals, while boys acquired proficiency in technical pursuits that would enable them to command good wages as men and to provide comfortable homes for their families.[28]

In working with children from preschool age through adolescence, Corwin hoped to create a better environment for future generations of adults and to eliminate, or at least to modify, the evils associated with saloons and drunkenness in the mining communities. Drinking, he insisted, was the major social problem among coal miners. Its cause, he believed, was the absence of economic opportunity and, frequently, conditions that prevailed in the homes. The remedy, therefore, was increased vocational and technical skills for men and a broader knowledge of domestic science for women.

The Sociological Department viewed saloons as "poor men's clubs" which, in the absence of alternative facilities, flourished as social centers in mining towns. In order to discourage patronage of local bars, the company established reading rooms, recreational halls, and clubs designed to perform a similar function without the excesses of drinking and the vices associated with the consumption of alcohol.

Reading rooms contained not only newspapers and periodicals in several languages, but in addition provided smoking and game rooms in which employees could relax at the close of the day's labor. No liquor was permitted on the premises, and gambling was strictly regulated. The first of these facilities had been opened in 1900 over the Colorado Supply Company's new retail store adjacent to the steel works. A similar unit was established in Villa Grove for workmen from the Orient Mine.[29]

Recreation halls were located at the Minnequa Hospital, in Starkville, and at Sunrise. Harmony Hall, in the camp near Trinidad, contained four rooms, one of which housed the kindergarten, another was the recreation center, and two smaller ones were equipped as a kitchen and wardrobe. By removing a folding partition, the two larger rooms were transformed into a hall for lectures, concerts, and dances. Sociological Hall in the Wyoming iron-mining town consisted of one room with a stage, as well as an alcove that served as the community library.[30]

The Sociological Department attempted in various ways to control the consumption of alcohol. In some camps where the company owned all of the property, saloons were prohibited. Confronted with that situation, miners who viewed drinking as an inviolable right and privilege resorted to bootlegging; the result was that drunkenness was frequently more of a problem in "dry camps" than where the sale of liquor was legal. Another experiment took the form of the regulated saloon that operated under terms of an agreement between the proprietor and the corporation. The penalty for permitting excessive consumption was removal of the saloonkeeper and his establishment from the community.[31]

Still another innovation was the "soft drink club." In at least one town, Floresta, a building was equipped by the Sociological Department with a reading room, billiard and pool tables, and a bar that served only coffee, chocolate, or bouillon. That facility was designed to serve the whole family rather than only the male wage-earner.[32]

In many of the newer camps, the solution to the drinking problem was the men's club that contained a "hard drink" bar, the operation of which was closely supervised in the interest of the company and the community. At Coalbasin, Redstone, and Primero, these facilities were erected by the company and administered by the employees, and they were the only sources of alcoholic beverages in the towns.

The first of these clubs was opened in Coalbasin where, in the summer of 1901, "a glorious" drinking spree had temporarily incapacitated most of the adult males. Although officially a "dry camp" at that time, some residents, in anticipation of the Fourth of July, smuggled into town eight barrels of beer, four kegs of whiskey, and a large amount of wine. In the aftermath of the celebration, work was suspended for several days at the mine which supplied coal to the ovens at Redstone.[33]

Reacting to that incident, the Sociological Department decided that Coalbasin was the proper place to test a new method for curbing the consumption of alcohol. It erected a clubhouse, one feature of which was a bar where drinks were sold under well-defined regulations. The one-story structure also contained game, billiard and pool, and reading rooms. The bar was plainly furnished in a conscious effort to avoid making it attractive, and there was no display of bottles or advertisements.[34]

The house rules, written in English, German, and Italian, were prominently displayed and enforced:[35]

1. The Club House will be open for the use of members from 9 a.m. to 10 p.m. daily except Saturdays, when it will remain open until 11 p.m.
2. Members whose occupations are such as to require special working clothes are requested not to remain in the Club Rooms in their working clothes.
3. No credit will be given to members or visitors. All charges must be paid at the time they are incurred.

4. No gambling will be allowed in the Club, but playing of cards for small stakes will be permitted, the stakes in no event to exceed the following limits:
   Poker — Penny ante and twenty-five cent limit.
   Billiards — 25 cents per cue.
   Pool — 10 cents per cue.
5. Women or children residing in or near Coalbasin will not be allowed to visit the Club Rooms except at such times as may be specified by the Board of Directors.
6. Strangers, including women and children, will be permitted to visit the Club Rooms for purpose of inspection between 9 a.m. and 5 p.m. except Sundays and holidays if provided with a permit from the Board of Directors.
7. No books or papers shall be taken from the Club Rooms.
8. Members will be charged for any damage done to the furniture or fixtures of the Club due to their carelessness or design.
9. No subscription paper shall be circulated nor any article exposed for sale in the Club House without the authority of the Board of Directors.
10. Notices shall not be posted on the bulletin board except upon the authority of the Board of Directors.
11. All talking in the reading room is prohibited.
12. No member shall use the billiard or pool tables for more than three consecutive games to the exclusion of others desiring to play.

One other rule was enforced in all clubs. No member or visitor was allowed to purchase a drink for any other person. The "no treating" regulation was designed to discourage excessive consumption of beer, wine, and liquor. Although frequently violated, that ban, according to Dr. Corwin, made the clubs an effective instrument for curbing drunkenness.[36]

The Coalbasin experiment was adopted for John C. Osgood's model village in the Crystal River Valley. The Redstone Club was incorporated for social purposes and governed by a 13-man board of directors composed of employees. Active members paid an initiation fee of $1.00 and monthly dues of $.50.

A two-story structure contained facilities designed to encourage the workingman to rest or seek diversions at the end of the day. On the first level, there was a spacious lounge and bar, flanked by a billiard and pool room and a reading area amply stocked with newspapers in English, Italian, and Slavonic. The upper floor was a theater, complete with stage and dressing rooms, scenery and drop curtain, electric lights, and an arch-illuminated stereopticon.[37]

Although ladies were not entitled to membership, the Redstone Club extended to them special privileges on the first and third evenings of each month. Then the wives, daughters, and friends of employees enjoyed full use of the rooms for euchre, billiards and pool, or other activities to the accompaniment of instrumental music. Only light refreshments were served on those occasions.[38]

Recreation halls and club houses also performed a variety of educational and social functions for the communities. They frequently housed the Sociological Department's circulating libraries and art collections. Lectures, stereopticon lantern shows, concerts, and plays were scheduled at regular intervals; dances, the universal pastime in all mining camps, occurred weekly or fortnightly.

In its quest for social betterment, the Sociological Department stressed improvements in the quality of miners' homes. The fuel and iron company, or its predecessors, had provided some dwellings for its employees for almost three decades, but little attention had been given to the matter of comfort or appearance of the structures. In many communities, residents had erected their own homes, often on the corporation's property, according to their own ideas of sanitation and beauty. These were a mixed lot, varying from neat, practical frame units to log cabins, adobes, dugouts, and shacks.[39]

When the new towns of Primero, Segundo, Tercio, and Cuarto were organized in the Purgatoire Valley west of Trinidad, the Sociological Department, in cooperation with the Fuel Department, designed homes that were comfortable and attractive as well as sanitary. Dwellings with from three to five rooms were erected on streets laid out on a rectangular pattern in each community. At the same time, pipes were installed to carry water from central reservoirs to all parts of the towns for domestic consumption and fire protection.[40]

At the four camps, more than 500 homes, mostly three- and four-room units, were constructed. They were rented to employees for $2.00 per room a month, a standard assessment throughout the camps. When available, electricity was furnished to residents at a cost of $.35 an outlet per month.

At the same time, Redstone was planned as a model village. At first, employees erected huts and shacks in the bottoms west of the ovens. These were torn down in 1902, after an auxiliary corporation, the Redstone Improvement Company, completed homes on the east side of the river. They ranged in size from three to five rooms, and were similar to units in other camps except that each of the larger ones contained a bathroom and rented for $18.50 per month, considerably more than the rate elsewhere.[41]

The houses in Redstone varied in design and color, and they were deliberately planned to blend with the natural beauty of the valley. One cottage was artistically and inexpensively furnished as a standard for other homes. Running water, sewers, and electricity, luxuries in other camps, were common to all units.

The community contained as well the Redstone Inn, a boarding house for single men, a club house, theater, bandstand, school, and library. South of the town and east of the river, John C. Osgood built a palatial private mansion. This accounted for the special interest he displayed toward the community and its inhabitants.

In Pueblo, where houses were readily available, the company did not rent homes to its employees. In 1900, a subsidiary enterprise, the Minnequa Town Company, in which Dr. Corwin was a director, built dwellings and other structures for the steel corporation. In three years, approximately 300 modern units, mostly of four and five rooms, were erected in Minnequa Heights, a subdivision west of the Minnequa Works. These houses were sold on contracts that required relatively small monthly payments.[42]

In order to encourage the work of social improvement throughout the company's scattered properties, the Sociological Department, from 1901 to 1904, published *Camp and Plant*. At a cost of $1.00 a year, subscribers, most of whom were employees, received an illustrated weekly (later monthly) magazine, the primary purpose of which was "to bring the various coal, coke and iron camps and works closer together; to help develop a stronger esprit de corps among employees . . . , and to furnish a medium through which the Sociological and Medical Departments can reach the people." Issues contained news of individual and community activities, including the secular and religious customs of ethnic groups in the mining camps.[43] There were also reports of social and industrial developments; articles, often in German, Italian, Spanish, and Slavonic, on sociology, hygiene and sanitation, domestic science, and education; as well as short stories, anecdotes, and original contributions from readers.

Camp and Plant sought to stimulate a constructive attitude toward employment. On the assumption that many men regarded their daily tasks as monotonous, editor Lewis stressed that work in the varied operations of the West's largest industry was exciting and meaningful. The magazine described the various properties and detailed the histories of mines and mills. Some articles recounted the processes of mining coal and iron ore, the production of coke, and the steps by which raw materials were transformed into iron and steel, and the methods by which those metals were fabricated into finished products at Laramie and Pueblo.[44]

The Sociological Department pursued its goal of social improvement until 1915. That year, following an unusually disruptive coal miners' strike, the Colorado Fuel and Iron Company, at the direction of John D. Rockefeller, Jr., adopted an Industrial Representation Plan as a means of restoring harmony between labor and management. At that time, the industrial department of the Young Men's Christian Association assumed the task of administering all social and educational programs at the steel works and in the mining camps.[45]

John Lathrop Jerome (1854-1903). A graduate of Hamilton College, New York, Jerome served as principal of public schools in Central City for a time before opening a law office in Denver. Jerome, along with Alfred C. Cass, Julian A. Kebler, and John A. Osgood, were known as the "Iowa Group," industrialists who opened Colorado's vast fuel resources and formed, in 1892, the Colorado Fuel and Iron Co. Courtesy of The State Historical Society of Colorado.

John Cleveland Osgood (1857-1926). A native of Brooklyn, Osgood was a successful coal operator in Iowa before moving to Colorado in 1882. He organized the Colorado Fuel Co., which merged in 1892 with the Colorado Coal and Iron Co. to form the Colorado Fuel and Iron Co. Osgood was that firm's chief executive officer from 1892 to 1903.

12

Osgood's Fight to Retain Control, 1901-1902

In early April 1901, Denver and Pueblo newspapers reported that the recently formed United States Steel Corporation had absorbed the Colorado Fuel and Iron Company. Even though J. A. Kebler issued a denial, rumors persisted that the "steel trust" had acquired the western enterprise in order to dominate the Pacific Coast market for iron and steel products, and to obtain coal cheaply for railway systems under its control.[1]

More than a year later, Judge Elbert H. Gary publicly admitted that the United States Steel Corporation, of which he was chairman of the board of directors, had attempted to purchase the Colorado company. "They made us a counter-offer," he said, "and neither proposition was satisfactory." The negotiations were not renewed.[2]

That story was essentially correct, but it omitted all of the details. In fact, prior to June 10, 1901, John W. Gates, who had recently acquired a large interest in the Colorado Fuel and Iron Company, had discussed with a member of J. Pierpont Morgan's New York banking house the question of a sale. The price asked in equivalent United States Steel Corporation stock was $175 for each of the western enterprise's outstanding shares, a figure that seemed unreasonable since they were selling substantially below that amount.[3]

Also in June 1901, Charles Schwab, president of the giant steel firm, talked to Osgood about the possibility of buying the Pueblo company. The Coloradoan insisted that Morgan "could not get the property on any better terms than one share each preferred and common United States Steel," or at current prices about $145 per share. Schwab considered that to be too high but agreed that the acquisition was desirable.[4] Meanwhile, Gates conferred with Morgan while both men were vacationing in England, and rumors of an impending purchase sent the price of the western firm's common shares to an all-time record high.[5]

The United States Steel Corporation, on October 10, 1901, formally offered to purchase control of the Colorado Fuel and Iron Company at an equivalent price of $105 per share. Gates and Osgood, the two largest stockholders, wanted to sell, but the latter refused to do so for less than $125.[6] Negotiations were finally terminated four months later. Because of his associate's resistance to the proposed transaction, Gates promised to oust him from control of the Pueblo enterprise's management.[7]

An upward surge in Colorado Fuel and Iron Company stock on the New York market had been attributed by "insiders" to Morgan's attempt to gain control of the western firm. In fact, most of the increase was caused by Gates and his associates who bought large quantities of shares in anticipation of big profits from the sale of the Pueblo steel works to the United States Steel Corporation.[8]

In attempting to gain a voice in the affairs of the Colorado Fuel and Iron Company, Gates set in motion a spectacular rise in the market value of stock. During the first three months of 1901, it had fluctuated from a low of 41 to a high of 66 with a small volume of sales. Early in April, the price shot upward. Before the end of the month, the Pueblo firm's shares surpassed par for the first time, and subsequently they went even higher.[9]

With the support of John Mitchell and James A. Blair, Gates claimed enough shares to control the board of directors. However, at that time, he professed a desire to work with Osgood and to leave him in charge of the enterprise. Gates did insist on further expansion of the steel works at Pueblo.[10]

That spring most of Osgood's friends sold

their fuel company stock. After having held it for several years at from one-third to one-half of par value, they could not resist the opportunity to sell at a handsome profit. However, Osgood was unable to take advantage of the favorable market without running the risk of losing the company. "I think they might have been a little more decent," he complained to Jerome, "as they must have known that they were simply selfishly taking their profits and letting me hold the bag."[11]

Although he lost substantial support, Osgood claimed a majority of the stock. At the close of April, he personally held 40,000 shares, controlled an additional 60,000 in the treasury of the Colorado Finance and Construction Company, and counted on the owners of 20,000, mostly the holders of preferred shares. He needed only 115,000 to retain the management.[12]

In the final week of April 1901, the contest was temporarily suspended when Gates, with Osgood's approval, made an agreement with E. H. Harriman and George J. Gould for the long-term management of the fuel company. Gates and Osgood were to contribute equally to a block of 50,000 shares which the railroad magnates were to purchase at less than the market price. That stock, together with an additional 76,000 shares from the same parties, was to be placed in a trust administered by the four men. They were to organize a new board of directors, with Osgood naming two members, Gates one, and Harriman and Gould one. The fifth position was to be occupied by Paul Morton, an executive with the Santa Fe Railway. In that way, the Colorado capitalist seemed assured of retaining control of the enterprise.[13]

The trustees planned to put whatever capital was necessary into the fuel company. New stock was to be issued at par, with each of the four parties taking an equal amount. It was also tentatively agreed that with the proceeds from new shares the enterprise should acquire the assets of the Victor Fuel Company, Northern Coal and Coke Company, American Fuel Company, and possibly the coal properties of the Union Pacific, Denver and Rio Grande, and Rio Grande Western railroads. In that way, a single corporation would control the fuel trade in Colorado, Wyoming, Utah, and New Mexico Territory.[14]

That scheme collapsed in early May. Harriman withdrew following a sharp break in the market price of Union Pacific stock. He had, according to Gates, exhausted his resources on the railroad and had no money for other ventures. Any hope of reviving the trust plan ended when E. P. Ripley, president of the Atchison, Topeka and Santa Fe, refused to allow Morton to serve on the fuel company's board of directors. The prospective trustees were unable to agree on a substitute, and negotiations were terminated.[15]

At that time, Osgood was deeply involved in discussions with Mitchell's Illinois Trust and Savings Bank and with Blair and Company about the possibility of issuing debenture bonds. The Coloradoan hoped to sell a total of $15 million, two-thirds of which were to be marketed immediately to underwrite another round of improvements at the steel works.

In the course of negotiations, Rodney Dennis, a member of Blair and Company, informed Osgood that in order to reassure the bankers who proposed to accept the bonds, it was essential that Mitchell and Blair be elected to the fuel company's board of directors. When Osgood agreed to this, Mitchell refused to serve unless permitted to name a majority of the body.[16]

On the evening of May 9, during the Northern Pacific Panic on Wall Street, Osgood met with Gates, Mitchell, and their attorneys at the Waldorf Astoria. The Chicago financiers demanded the right to name a majority of the Colorado company's board of directors. When Osgood refused, they subjected him to various pressures, including the possibility of abandoning the projected issue of debenture bonds. They also threatened to throw their fuel company stock on the market, which would cause a precipitous decline that might be disastrous to Osgood and the corporation.

When the Coloradoan continued to resist, Gates and Mitchell turned to a number of his brokers who had been invited to the meeting. They implored Osgood to concede. If the Chicagoans broke the market, the security dealers would be ruined along with Osgood because of his large transactions in margined stocks which they carried on their books.[17]

Reluctantly, Osgood yielded to the demand. He made an oral agreement to allow Gates and Mitchell to name a majority of the directors, with the assurance, in return, that they would not change the existing management. The next day, the Illinois Trust and Savings Bank and Blair and Company agreed to market the first installment of the debenture bonds.[18]

Although Osgood viewed the concession as having been obtained under duress and therefore improper, he was not then in a position to repudiate it. He feared that if he did not abide by the agreement, Gates and Mitchell would carry out their threat to demoralize the market in the fuel company's stock.[19]

During the month of June, when the stock reached a high of 136, Osgood sold a large amount of his holdings in anticipation of buying them back at a lower price. In that way, he put his personal finances in good order and was able to purchase the 60,000 shares from the Colorado Finance and Construction Company, which strengthened his position for the forthcoming annual meeting of stockholders.[20]

That month, Osgood decided to ignore his agreement with Gates and Mitchell and to fight for control of the company. In an effort to bolster his position, he called for a special meeting of stockholders on June 13, at which time a plan to issue debenture bonds for financing further modernization and expansion of the steel works and fuel department was overwhelmingly endorsed.[21]

In a letter to Mitchell on August 10, less than two weeks prior to the regular annual meeting, Osgood repudiated the agreement of May 9. He announced that he planned to secure a majority of his own, and that if unsuccessful, he would withdraw from the management. However, he did agree to support the election of four directors to represent Gates and his associates.[22]

Nine days later, Gates capitulated. He submitted his proxies to Osgood, thereby assuring unanimous action at the annual meeting. On August 21, Gates, Mitchell, Blair, and John Lambert were elected to the board along with Osgood, Kebler, Cass, Jerome, Beaman, Sullivan, Grant, James, and Joseph Heimerdinger. The last-named director, selected for the first time that year, was Osgood's principal broker in New York.[23]

Although beaten, Gates did not give up the fight. The following year, he renewed the contest, which was fought in the state and federal courts. For a period of two months, while the outcome remained in doubt, each side exploited the local press in an attempt to influence uncommitted shareholders.

The contest opened formally on July 12, 1902, when Arthur J. Singer, a member of the New York brokerage firm of Harris, Gates, and Company, demanded of the Colorado Fuel and Iron Company's transfer agent and registrar, the Knickerbocker Trust and Atlantic Trust companies, respectively, the right to examine stock transfer and registry books. On the assumption that Singer sought information to advance a planned take-over of the western enterprise, Beaman, as secretary, refused to open the records for inspection.[24]

Singer then brought a mandamus suit in the New York Supreme Court to gain access to the documents. An order was rendered in his favor. However, by agreement with Knickerbocker Trust, he relinquished his right to examine the stock book in return for a certified list of current stockholders.[25]

Gates and his associates carried their legal battle to the West. On July 21, D. H. Dougan, a shareowner, asked to examine the stock book in the Denver office. When Beaman denied that there was such a record, Dougan filed suit in the District Court of Arapahoe County to compel the company and its officers to make available the desired data. The complaint charged that Osgood and his associates planned to disfranchise large numbers of stockholders at the forthcoming annual meeting. For reasons that were not revealed, the legal action was dismissed at the request of plaintiff prior to hearings on his petition.[26]

Colorado's statutes clearly required all corporations except railways and telegraph companies to maintain stock books in their principal places of business within the state. An amendment of April 14, 1893, to the corporation law stated that the record had to contain the names of all persons, alphabetically arranged, who were or had been stockholders within the past year, with their places of residence, numbers of shares owned, dates they acquired or sold stocks, the amounts actually paid in, and the proportion purchased in cash. All transfers had to to be recorded within 60 days, and the book was to be open for inspection by stockholders, the corporation's creditors or their representatives, and shareowners were entitled to make extracts from the records.[27]

On July 29, Singer, as a stockholder, asked to see the stock book kept by the company in Denver under the terms of Colorado law. When compliance was not immediate, he initiated another suit, this time in the state's District Court in Denver to require the corporation to prepare and maintain the record in question, and to make it available for inspection. The petition was dismissed on the grounds that it failed to show that the firm's officers had ordered the keeping of such a book, and that they had not been named defendants in the legal action.[28]

What had become almost a ritual was repeated on August 7, when Denver attorney William N. Vaile, a stockholder, wrote to Beaman asking to inspect the Colorado stock book. When the corporation's secretary again refused on the grounds that the book did not exist, Vaile brought suit in the state court to compel compliance with this request, alleging that the record was in fact in Beaman's office. One week later, the court ruled that the plaintiff had failed to substantiate the charge, and the petition was dismissed.[29]

The Colorado Fuel and Iron Company's board of directors on August 13 ordered the secretary to prepare the stock book required by Colorado law. Since there were many old records that contained 14,000 certificates requiring about 35,000 entries, all of which had to be checked against the transfer records in New York, it seemed unlikely that the task could be completed by August 20, the scheduled date for the annual gathering of stockholders.[30]

On the assumption that the record would not be available in time for the meeting, Gates, Mitchell, Blair, and Lambert asked that the eligibility of voters be determined by the New York transfer agent's certified list of stockholders, as of July 31, 1902. Since the board had ordered preparation of the requisite record, and because the certified list was not acceptable under Colorado law as a substitute for the stock book, no action was taken on the request by the minority directors.[31]

Meanwhile, Gates and his friends, according to reports that circulated in Denver, sought to strengthen their position by making an alliance with George J. Gould and Edwin Hawley, who were promised seats on the next board of directors. Because of their respective interests in the Denver and Rio Grande and the Colorado and Southern, it was assumed that the railroad magnates accepted the offer in order to protect the traffic both lines enjoyed as carriers of raw materials and finished products for the Colorado Fuel and Iron Company.[32]

At the same time, the group contending for control sought to discredit the existing management. Gates's supporters charged that the board of directors dominated by Osgood had refused to authorize a quarterly dividend in order to undermine the market price of the corporation's securities. They were guilty of stock-jobbing. In fact, according to those critics, earnings were more than sufficient to justify a distribution to shareholders, and the improvements at Pueblo, instigated by Gates and his associates, would be completed long before accumulated surpluses had been exhausted.[33]

In response, Osgood declared that the rival faction knew that the company's financial condition did not justify payment of a dividend because construction at the steel works had disrupted production and diminished profits. Work on the improvements had been halted for a month by a strike of bridge and structural-iron men against the contractors who were erecting the new mills, and there was a real danger that the labor disturbances might spread to the company's operating departments.[34]

As to the construction in progress, Osgood insisted that he and his associates had planned and raised the money for the new production units. The members of the opposing faction, he said, had done nothing more than to "subscribe for the debentures, which they did at a large profit to themselves."[35]

In the contest with Gates in the summer of 1902, Osgood was clearly the underdog. By that time, the Coloradoan had disposed of much of his stock, as had most of the friends and associates on whom he had always counted for perpetuating his control of the enterprise. Under the circumstances, Osgood had no choice but to stall for time while he sought to gain additional strength for the showdown in the annual meeting.

Two schemes were adopted to prevent Gates from gaining control of the fuel company. The first called for Osgood, as presiding officer at the forthcoming annual meeting, to reject the Chicago financier's proxies on the grounds that they had not been lawfully recorded in the Colorado stock book. In this way, the current management was assured of a large majority and could dominate the board of directors for another year.[36]

Another proposal, one that was not seriously considered at first, required that Osgood and his associates petition the state courts for an injunction enjoining the annual meeting. That way, the existing board would remain intact for an indefinite period of time.[37]

When John Jerome objected to both plans on the grounds that they were high-handed and eventually would be of no avail, Osgood insisted that he proposed only to do those things that were strictly legal. In addition to that, he argued, whatever action he adopted was justified by the numerous breaches of faith that Gates and his associates had committed in their recent relations with the company's management.[38]

On August 15, Mitchell, Blair, Gates, and Singer filed in the United States Circuit Court at Denver a bill in equity, making the fuel company and its directors defendants. The plaintiffs asserted that they owned or controlled 191,609 shares and that until the state court had ruled in Vaile's suit, they did not know that a stock book, which was required by law, was unavailable in Colorado. They also charged that the board of directors in a recent action had prevented some stockholders from participating in the next annual meeting. Therefore, they asked the court to issue a temporary injunction to protect their rights in the election of directors for the following year.[39]

On June 30, the board had clarified the meaning of the by-laws on the question of who should conduct annual meetings of stockholders. The Gates faction had contended that the presiding officer could be elected by a majority of the shares present. The board rejected that view and adopted an amendment that named the officers of the company, in order of official priorities, as chairmen. Only in the absence of an officer could the election come from the floor. At the same time, the chairman was given the right to appoint tellers and to employ a cumulative ballot in the election of directors, as sanctioned by Colorado statute.[40]

In answering the latest charges, attorneys for the fuel company stated that more than one-half, or approximately 100,000, of the shares claimed by Gates and his associates were invalid proxies which were sufficient, if voted, to defeat the interest of bona fide stockholders. The complainants, according to "The Joint and Several Answers of Said Defendants," had temporarily borrowed or purchased shares which were registered prior to the closing of the New York transfer books, then returned to their real owners. In that way, those who challenged the existing management gained the authority to vote a large amount of stock which they did not own or control. Furthermore, the holders of those proxies were in no way concerned "as to the welfare of the ultimate success of the company."[41]

On August 19, one day prior to the scheduled annual meeting, Judge John A. Riner refused to issue the temporary injunction requested by Gates and his friends. In denying that remedy, the court ruled that the meeting had to be held in accordance with Colorado statutes under which the by-laws in question were proper and valid.[42]

At this point, Osgood decided to seek a court order to block the convening of stockholders. This action was probably prompted by the opposition's emphasis in recent court actions on the nonexistence of a stock book in the Denver office. The record, then in preparation, was not available to determine the eligibility of shareowners in electing directors for the following year.

On the evening of August 19, Jerome, who was convinced that Osgood could only delay but not avoid loss of the company, decided to seek a last-minute compromise between the rival factions. With Osgood's knowledge, he met with Gates and Mitchell at the Denver Club to explain that opposition to their efforts to gain control of the board of directors arose principally from concern for the auxiliary corporations in which the current management had large interests. These enterprises were the Crystal River Railroad, Colorado and Wyoming Railway, Rocky Mountain Coal and Iron Company, Pueblo Realty Trust Company, Colorado Supply Company, Mountain Telegraph Company, Laramie Iron and Steel Company, Minnequa Cooperage Company, Rocky Mountain Timber Company, Minnequa Town Company, Redstone Improvement Company, and Steel Wheel and Wagon Company.[43]

When both parties agreed to discuss the future of the subordinate corporations, Jerome arranged for a meeting between Osgood and Gates's representative, Edward O. Wolcott. The talks failed to lead to an understanding. Osgood charged that Wolcott was overbearing and that he demanded humiliating concessions as a condition for purchase of the properties. Wolcott, on the other hand, insisted that Osgood, in order to "sustain his dignity," wanted to retain some power in the fuel company, or at the very least to place Gates in a "beaten position."[44]

On the morning of August 20, George F. Bartlett, a stockholder in the Colorado Fuel and Iron Company and secretary of the Victor Fuel Company, filed in the District Court, Arapahoe County, a petition for an injunction to prevent stockholders from conducting a regular annual meeting later that day. The plaintiff asserted that rival factions were seeking to gain control of the Pueblo corporation and that in the absence of a stock book it was impossible to determine the respective rights of the contenders. Therefore, if the election of directors were held as scheduled, it would result in the selection of two sets of officers to the detriment of the enterprise. Furthermore, because proper notice had not been published as required by law, the meeting would be illegal. Only when the stock book, then in preparation, had been completed, would it be possible to conduct the meeting without harmful controversy.[45]

In response to Bartlett's petition, Judge John L. Mullins issued a temporary injunction and ordered that it be served on stockholders who were about to convene at the Boston Building in Denver. They were restrained from conducting any business other than the adoption of a motion for adjournment.[46]

When stockholders met at the Colorado Fuel and Iron Company's general offices at three o'clock, the Sheriff of Arapahoe County read the writ of injunction of the assembly. Thereupon, Beaman presented a protest of 18 stockholders against the taking of any action on the grounds that the annual meeting was illegal because no newspaper notice had been published as required by law and because there had not been time to prepare the stock book which constituted "the only proper and legal evidence as to what persons are legally entitled to vote at any election of directors."[47] Osgood, as chairman, then announced that for those reasons, which he supported, and because of the injunction previously read, "there can be nothing done but to adjourn this meeting subject to further order of the court."[48]

One week later, Gates, Mitchell, and Lambert asked the federal court in Denver to accept jurisdiction over the Bartlett suit. They argued that the state court was subject to local influence which denied them assurance of an impartial hearing. On September 3, the higher court accepted the case, at which time the defendants filed a cross-bill of complaint against Bartlett, the Colorado Fuel and Iron Company, and its officers.[49]

On October 2, Judge Henry C. Caldwell rendered a decision in the Bartlett suit and the cross-complaint. The substitute annual meeting for the election of directors was scheduled for December 10, and the proceedings were placed under the direction of Judge Seymour D. Thompson of St. Louis as Special Master in Chancery. He was empowered to subpoena witnesses, take testimony, and examine all pertinent records in order to determine the eligibility of stockholders to participate in that election.[50] On the first day of the following month, the United States Circuit Court of Appeals denied to the company and its officers the right of appeal of Judge Caldwell's ruling to the United States Supreme Court.[51]

By that time, Gates and his friends had abandoned the fight to overthrow Osgood. They sold their Colorado Fuel and Iron Company stock, or at least a sizeable portion of it, in an effort to recover some of their investment. According to one estimate, they suffered an aggregate loss of nearly $7,000,000 because of the rapid depreciation of the shares following the adjournment of the regular annual meeting.[52]

Locally, no sympathy was expressed for the Chicagoan and his colleagues. The people, one newspaper asserted, are almost unanimous in sentiment that the Colorado Fuel and Iron Company should not pass into the hands of the "buccaneers of Wall Street."[53] The pastor of a Denver church described Gates as a "speculative free-booter" who sought to profit from the manipulations of corporate securities.[54]

With apparent glee, a Denver newspaper published an example of "flippant banter" from a New York daily:[55]

> Plunger Gates went away
> With much ado, a great display
> Of banners floating and trumpets sounding
> And triumph on his bosom pounding
> Plunger Gates in coming back
> And sighs and tears to mark his track
> In sweeping down on Eldorado
> He was trapped and skinned in Colorado.

Following the adjourned annual meeting, Osgood returned to New York to seek allies in his fight to retain control of the fuel and iron company. In late September, he opened negotiations with Edwin Hawley, E. B. Harriman, and George J. Gould, proposing that they join him in organizing a proxy committee. If that plan succeeded, Osgood was to nominate a majority of the directors and dominate the management of the corporation.[56]

Negotiations continued for about six weeks, encouraging Osgood to believe that with the assistance of the railroad leaders he could retain control of the western enterprise. However, on November 5, Hawley announced that he and his associates insisted on naming a majority of the board, and that Osgood had to be content with minority representation. When the Coloradoan refused, Hawley and Harriman, acting also for Gould, issued a circular that requested proxies for the annual meeting scheduled by the court. Osgood prepared and distributed a circular of his own, naming himself, Kebler, and Jerome as a proxy committee in behalf of the current management.[57]

The Harriman-Hawley-Gould alliance prompted Osgood to seek support from the Atchison, Topeka and Santa Fe, whose business might suffer if the Union Pacific-Southern Pacific, Colorado and Southern, and Denver and Rio Grande gained interests in the Colorado Fuel and Iron Company. Enroute to New York in late October, Osgood stopped in Chicago to persuade E. P. Ripley that the Santa Fe should purchase a large amount of the Pueblo firm's stock or else should make available a loan with which Osgood could secure more shares in his own name. Harriman, Hawley, and Gould controlled only 60,000 shares, or one-fourth of the outstanding stock, according to Osgood, and the rest, except for 90,000 in Osgood's name, were "floating on the market."[58] Ripley concluded that it was in the interest of his company to block if possible the acquisition of the Colorado industry by men who would be less friendly than the present management. Whatever else might be said of Osgood, he was loyal to his friends; if the Santa Fe aided him in retaining the enterprise, "it would be nearly the same thing as control for ourselves as far as the business goes."[59] However, the railway's executive committee refused to assist Osgood because there was no assurance that the expenditure of even $2,500,000 would secure control of the steel works.[60]

Therefore, the Santa Fe chose to rely on the magnitude of its business to obtain fair treatment in the event that the Colorado enterprise passed into the hands of other railroads. This appeared to be a realistic decision. If the fuel and iron company were "run on its merits as a manufacturing concern," it could not safely withdraw a large volume of traffic, because the Santa Fe was a valuable customer, and its good will was necessary.[61]

The railroad syndicate was short-lived. On November 15, Gould, who had been absent from New York when Hawley and Harriman requested proxies, disavowed that action. Six days later, he sent out a call for proxies, and his former associates made a separate appeal of their own.[62]

By that time, Osgood and Gould had formed an alliance. Each man solicited proxies, but they agreed to cooperate in naming a majority of the directors. They hoped to gain nine members, four each plus one who was acceptable to both parties. The management was to remain unchanged.[63]

The evening before the scheduled substitute annual meeting, the three proxy committees assembled in the Colorado Fuel and Iron Company's general offices. That meeting, under the supervision of Judge Thompson, produced an understanding for a division of the board of directors that was to be elected the following day. Each faction agreed to name four men, and by common consent, James H. Hyde, vice-president of the Equitable Life Assurance Company of New York, was to occupy the other seat.[64]

The next morning, stockholders gathered in President Kebler's office for the annual meeting. Although scheduled to commence at ten o'clock, it was delayed while D. C. Beauman and his assistants drafted requisite documents. "To do this now," Osgood explained, "will obviate any delay in the meeting."[65] Finally, at three minutes past eleven, the chairman called the group to order. After reading the notice which had appeared in Denver newspapers ten days earlier, the secretary announced that the suit in the federal court instituted by John W. Gates had been dismissed, and that Judge Thompson for that reason was a spectator without authority at the proceedings in progress.

According to the prearranged plan, all shares were voted for each of the candidates nominated by the three proxy committees. Osgood, Kebler, Cass, and Jerome were the only men who retained seats from the previous board. Gould was elected, together with C. H. Butler, partner and principal counsel for George P. Butler and Brother of New York; John H. McClement, a member of the same firm; and J. M. Herbert, president of the Denver and Rio Grande Railroad. Hawley and Harriman were chosen, plus H. E. Huntington, president of the Southern Pacific Railroad, and Frank Trumbull, president of the Colorado and Southern and the Colorado Midland railways. The thirteenth member was James H. Hyde.[66]

The annual meeting lasted less than fifteen minutes. Before adjourning, Osgood stated that the new board would convene after lunch to select officers for the next year.

That afternoon, the directors reelected Osgood chairman of the board and named him to the executive committee along with Kebler, Cass, McClement, and Gould. The officers from the previous year, with one exception, were reappointed. Kebler was named president, Cass first vice-president, Jerome third vice-president and treasurer, and Beaman secretary. McClement, as Gould's representative, was named second vice-president to succeed Dennis Sullivan. C. S. Robinson remained general manager of the iron and steel department, and John T. Kebler continued in a similar capacity in the fuel department.[67]

The events of December 10 were hailed in Pueblo and in the mining camps as a great victory for Osgood and as assurance of continuity in the management of the fuel and iron company. At Engleville, a huge bonfire was ignited on a hill overlooking the town, and 200 pounds of dynamite "awoke the surrounding canons."[68] Because of the intensity of the celebrations, nearly all of the mines remained closed the following day.[69]

Osgood quickly issued a victory statement in which he assured employees and the public that the firm's policies would not be changed. The corporation, he affirmed, will "go forward in developing the fuel and iron interests of the Rocky Mountain region . . . in harmony with the railway and other interests of the West as in the past."[70] All members of the new board of directors were determined to "pull together" and would allow nothing to interfere with the progress of the company.

13
The End of the Osgood Era, 1902-1903

Within less than a year after the stockholders' meeting of 1902, Osgood was forced to surrender control of the Colorado Fuel and Iron Company. By terms of a financial reorganization in mid-1903, the enterprise averted insolvency, but the direction of its affairs passed into the hands of eastern capitalists. The "Iowa Group," after having struggled for more than a decade to sustain heavy industry in the West, relinquished direction of the Pueblo corporation and its ancillary enterprises.

On the assumption that all major improvements would be completed in 1902, Osgood had anticipated for that year earnings of more than $3,000,000, enough to meet all expenses and to pay dividends on preferred and capital stock.[1] However, construction proceeded slowly and seriously disrupted normal production, leading to a sharp decline in income.[2] Furthermore, the company was plagued by a series of labor disputes that interrupted operations in several departments at Pueblo and caused additional delays in the completion of new furnaces and mills.[3]

In March 1902, Jerome warned Osgood of an impending financial crisis. The debenture bonds issued the previous years were not sufficient to pay for improvements in progress and to provide adequate working capital. Additional securities, authorized but unissued, had been allocated for the acquisition of coal and iron properties, water resources, and land for future expansion at the steel works. Even if the whole of that amount were devoted to current projects, Jerome argued, it would not be enough to avert a critical situation.[4] Convinced that there was no reason for alarm, Osgood insisted on going ahead with plans for enlarging the company's output of iron and steel products. When, on May 2, 1902, the board of directors authorized the start of construction on Furnace F at an estimated cost of $850,000, Jerome objected strenuously on the grounds that delays in current programs threatened financial disaster. Osgood replied that there was no danger. Long before the last of the income from the debenture bonds had been exhausted, he planned to raise additional funds by placing mortgages on the properties of the Crystal River and the Colorado and Wyoming railways.[5]

On June 30, the directors authorized the issue and sale of the remaining $5,000,000 in debentures, which were sent to the Illinois Trust and Savings Bank for certification. Three days later, Mitchell refused to take appropriate action, arguing that the agreement of August 1, 1901, did not permit the release of that large an amount at one time.[6] Jerome warned the Chicago banker that unless he acted quickly the fuel and iron company could be forced into receivership. After wrangling for three weeks, Mitchell agreed to approve three-fifths of the proposed issue, with the balance to be released in three installments during the next six months.[7] On July 30, when Osgood applied to the New York Stock Exchange for listing of the bonds, a protest from Harris, Gates, and Company caused another delay.[8] The debentures were finally listed and sold August 13. In the interim, Jerome borrowed funds in Denver and Chicago.

In December 1902, Osgood finally recognized the possibility of a financial crisis. The corporation needed at least $3,000,000 more than what it could raise from resources at hand, including the railway bonds. Jerome predicted that unless additional funds were found the enterprise would be in default by mid-March.[9]

The Colorado Fuel and Iron Company, by borrowing from Denver banks and discounting all available notes, paid some $500,000 in obligations due March 1, 1903. However, by that date, the crisis had reached serious proportions. Two days later, Osgood returned to New York to seek financial assistance from George J. Gould and John D. Rockefeller and to impress upon them the critical state of the western enterprise's condition.[10]

Commencing in mid-March, Osgood borrowed over a period of three months from Gould and Rockefeller, through George P. Butler and Company, a total of $2,798,000 for the fuel and iron company. Sixteen loans were secured by the stocks and bonds of the Colorado and Wyoming Railway.[11]

In early May, J. H. McClement spent several days in the Denver offices making an examination of financial conditions. Afterward, he estimated that the corporation had to have between $7,000,000 and $10,000,000 to meet pressing needs. Jerome suggested that the only assets for raising so large a sum were the unmortgaged steel works water system and Wyoming iron mines, plus equities in certain other properties. The two men also discussed at that time the possibility of increasing the company's capital stock and marketing it at less than par through an enterprise similar to the Colorado Finance and Construction Company.[12]

At about that time, Osgood admitted that he could not prevent the financial collapse of the company. He asked Jerome, in the event of bankruptcy, to protect the loans outstanding from Denver banks and the Commercial National Bank of Chicago. He also directed him to seek from the legal department advice as to the possibility of having Kebler named receiver of the enterprise. Jerome reported that the bondholders would have a voice in naming the court-appointed administrator and, since the Illinois Trust and Savings Bank held a large percentage of the bonds, it seemed unlikely that Mitchell would accept in that role one of Osgood's close friends and business associates.[13]

On June 15, the company failed to pay $650,000 due in New York. When the notes went to protest, the price of the corporation's shares declined sharply and creditors talked of obtaining a receivership. The obligations were paid in full the following day, but the experience had severely damaged public faith in Osgood's enterprise.

Officially, the default had occurred because of a misunderstanding between an officer and employee, caused by an error in telegraphic communication from New York to Denver. Off the record, the company failed to pay the notes because it was short $200,000 of the amount due.[14]

On June 23, Osgood agreed to step out of the company if Gould and Rockefeller would save it from insolvency. That proposition was immediately accepted. On the following day, Osgood, Kebler, Cass, and Jerome resigned as directors and as officers of the enterprise. However, Jerome, for the sake of appearances, agreed to remain as treasurer until the financial crisis had passed. McClement was elevated to the chairmanship; and Frederick T. Gates, E. Parmalee Prentice, and John D. Rockefeller, Jr., were named to vacancies on the board.[15] With those appointments, the Rockefeller interests emerged as the dominant influence in the corporation's affairs.

The fuel and iron company was without funds — and its obligations amounted to nearly $7,000,000. Unless it obtained immediate relief, the enterprise would be forced into receivership and reorganization.[16]

Unable to borrow from banks or to sell securities in the market, the company turned at that critical moment to its largest stockholders, Gould and John D. Rockefeller, Jr. Through an agent, Edward V. Cary, they agreed, by terms of a contract dated July 15, 1903, to purchase for the sum of $5,400,000 certain assets that included the capital stock of the Colorado and Wyoming Railway Company, the Crystal River Railroad Company, and the Rocky Mountain Coal and Iron Company, in addition to the bonds and notes of the railroads and the lease and purchase option on the Wyoming iron mines. Credited against that amount were the previous loans plus interest, aggregating $3,053,502. Cary also advanced $1,000,000 at six percent interest per annum to meet current expenses and payrolls.[17]

The annual meeting of stockholders on August 19, 1903, revealed that the Gould-Rockefeller alliance dominated the corporation. Together the two men controlled 9 of 13 seats on the directorate. Elected that day were Osgood, Kebler, Hawley, Harriman, Gould, McClement, Hyde, Prentice, F. T. Gates, J. D. Rockefeller, Jr., Ben Nicoll, E. W. Oglebay, and Frank J. Hearne.[18] Although Kebler, for reasons of health, had resigned from the presidency, he and Osgood agreed to serve on the board and to act in an advisory capacity to Hearne, who was named chairman of the board of directors and president of the corporation.[19]

Formerly chief executive officer of the National Tube Company of Pittsburgh, Hearne had remained with that enterprise as general manager after it was absorbed in 1901 by the United States Steel Corporation. He resigned to assume charge of the Colorado Fuel and Iron Company, whose directors on September 2, 1903, formally ratified a five-year contract that gave him an annual salary of $50,000 and personal control of the corporation.[20]

The stockholders' annual meeting of 1903 marked the close of an era in the history of the fuel and iron company. Control formally passed from the men who had organized the enterprise only 11 years earlier. With the ascendancy of the Gould-Rockefeller interests, Osgood and his associates salvaged only a small part of the large industrial empire they had created. The most valuable of the auxiliary companies were lost. At the annual meetings of the several enterprises on August 19, 1903, majorities of Gould-Rockefeller men were named to the boards of the Crystal River Railroad Company, Colorado and Wyoming Railway Company, Colorado Supply Company, Mountain Telegraph Company, Laramie Iron and Steel Company, Minnequa Cooperage Company, Rocky Mountain Timber Company, and Minnequa Town Company.[21] The former management group retained possession of the Crystal River Land and Improvement Company, Redstone Improvement Company, and Steel Wheel and Wagon Company, all of which had derived their chief value from closing working relationships with the fuel and iron enterprise.

The unsuccessful struggle waged by the "Iowa Group" to retain control of the Colorado Fuel and Iron Company exacted a heavy toll. Within a period of five months, the three men on whom Osgood had for many years depended for the efficient management of his enterprises passed away. Alfred C. Cass died July 4, 1903, at Redstone, following a general breakdown attributed to overwork. Julian A. Kebler, whose health had deteriorated alarmingly, suffered a cerebral hemorrhage on November 20, 1903. The following day, John L. Jerome, distraught over personal finances, was the victim of a self-inflicted gunshot wound while at his summer home in Buffalo Park, west of Denver.[22]

Although deprived of his closest friends and business associates, Osgood quickly turned his entrepreneurial talents to the task of regaining a commanding position in the western fuel trade in competition with the powerful Colorado Fuel and Iron Company. He remained a director of the Pueblo firm until September 1904, participating in the financial reorganization of that enterprise, while also administering the Victor Fuel and American Fuel companies, ventures which he had acquired in recent years through the Colorado Finance and Construction Company.[23] The Victor corporation, operator of mines in Fremont, Huerfano, and Las Animas counties, was Colorado's second-ranking producer of coal and coke.[24] The American Fuel Company, with mines in the Gallup district of New Mexico Territory, was a major supplier of fuel to railroads and commercial consumers in the Southwest.[25]

The change in the Colorado Fuel and Iron Company's administration afforded Osgood an opportunity to advance his position in the fuel trade at the expense of the corporation he had organized and directed for many years. Reacting to the ascendancy of rival railroad interests, the Atchison, Topeka and Santa Fe Railway renegotiated its contract with the Pueblo firm for the operation of coal properties in Fremont County. A supplemental agreement, effective August 20, 1904, reaffirmed the lessee's control of the Rockvale and Brookside mines, plus adjacent lands, but terminated leases on other tracts.[26] The latter were turned over to the Victor Fuel Company, from which the Santa Fe agreed to purchase one-half of its daily requirements of Canon coal.[27]

Anticipating a reduction in business with

the steel works, Ripley sought to enlarge his carrier's coal and coke traffic in Colorado. He decided that the Pueblo firm should not retain absolute control of the railway's coal properties when Osgood, "a very large operator in our territory, and one whose influence will always be used in our behalf," was willing to open new mines, the products of which would travel to market over the Santa Fe's tracks.[28] Therefore, Ripley leased to the Victor company the so-called "south field," in which the firm immediately developed the Radiant Mine, located four miles south of Coal Creek, for production of Canon coal.[29] Ripley believed that the demand for that product had increased faster than the supply in recent years, and that the Colorado Fuel and Iron Company, because of its operations in other districts throughout the state, had been unwilling to enlarge its output in Fremont County.[30]

In addition to the Radiant slope, the Victor Fuel Company worked in 1904 a total of nine mines in Colorado. The largest were in the Trinidad district where the Hasting, Delagua, and Gray Creek properties also featured beehive ovens for the manufacture of coke. The Chandler and Maitland mines, in Fremont and Huerfano counties, respectively, supplied fuel for railroads and other commercial users, along with a large domestic market.[31]

The Colorado and New Mexico operations were combined in 1909 as the Victor-American Fuel Company. Incorporated in the state of Maine with an authorized capital of $7,500,000, the enterprise owned or leased more than 50,000 acres that contained an estimated 500,000,000 tons of marketable coal. The corporation also controlled the Colorado and Southeastern Railroad, linking mines at Hastings and Delagua with regional carriers, and the Western Stores Company, whose merchandise outlets were located in mining camps in Fremont, Huerfano, and Las Animas counties, and also the Gallup district.[32]

Although John C. Osgood lived in New York where he preferred to conduct business affairs, he maintained for several years his official residence at Cleveholm, a baronial mansion at Redstone. After the Colorado Fuel and Iron Company suspended coking operations in the Crystal River Valley in 1909, that model industrial community was abandoned and left to decay. Osgood closed his home in 1913, after which he spent much of his time traveling in Europe.[33]

Twelve years later he returned to Redstone in poor health to spend his remaining days in the natural beauty of the Crystal Valley. Stricken with abdominal cancer, he died at Cleveholm on January 4, 1926.[34] His passing was largely unnoticed in Colorado where only a quarter of a century earlier he had been widely hailed as the state's leading industrial capitalist.

Epilogue: CF&I in the Twentieth Century

The Colorado Fuel and Iron Company entered the twentieth century stronger in terms of physical facilities than it had been at any time in the past. The corporation's predominance in the fuel trade was firmly established and seemingly unshakable. Furthermore, metallurgical production had been favorably advanced by a reconstruction and expansion program that increased the capacity of the Minnequa Works, lowered production costs, and yielded a greater variety of products for the western market.

However, there were problems that had to be resolved if the Pueblo firm hoped to achieve a condition of lasting prosperity. The fuel trade, for example, was threatened by two alarming developments. The coke market had been undermined by the decline of precious-metal mining and the decay of the smelting industry. Even more dangerous was competition from petroleum and natural gas in the market traditionally commanded by coal. Railroads and other commercial consumers were turning to the rival fuels, and eventually even the majority of domestic users adopted the cleaner, more convenient fuels.

Precious-metal mining had suffered a reversal in 1893 when the United States abandoned bimetallism. Colorado's output of silver fell from $160,000,000 in the decade of the nineties to $72,000,000 in the first ten years of the twentieth century.[1] The decline continued after 1910, with the result that most smelters of lead- and copper-base ores significantly reduced output or ceased operations.

The decline of the smelting industry had a dramatic impact on the Colorado Fuel and Iron Company, the region's largest manufacturer and distributor of coke. The El Moro ovens were abandoned in 1908; in the following two years, the plants at Redstone and Cardiff were closed.[2] Curtailment of coke production adversely affected mining operations in Las Animas County and in the mountain district. The remainder of the beehive and stack-oven plants, with one exception, suspended operations in 1918. Only Segundo continued to supply what remained of the smelter market and acted as a reserve for the steel works. When these ovens burned in 1929, they were not rebuilt.[3]

The company centralized its coking operations at the Minnequa Works, where by-product ovens supplied the steel plant, in addition to generating a variety of chemicals for the regional market. Two batteries of ovens commenced making coke in July 1918.[4] These were augmented by two more batteries over a period of little more than two decades, raising the number of ovens to 206. Coking gas yielded benzol, naphtha, and other products, in addition to serving as fuel for the ovens, rolling-mill soaking pits, and boilers. Unsalable chemical residues were consumed as fuel in the open-hearth department.[5]

After peaking in 1912, coal production declined in Colorado until World War I, when output soared to record levels. During the years from 1904 to 1920, the Colorado Fuel and Iron Company remained the leader in the fuel trade,[6] opening or acquiring by purchase 14 mines and closing 12.

In the southern field, long the center of fuel operations, the company shifted production from coking to steam and domestic coals. In the Trinidad district, new mines included Quinto (1905), Frederick (1907), Morley (1907), Cornell (1909), and Toller (1918). The latter was purchased from the Cedar Hill Coal and Coke Company and Giacamo Toller. The Quinto, Cornell, and Tercio mines were closed in the years from 1907 to 1915. Near Walsenburg, the Ideal, Lester, and McNally mines were purchased prior to 1912. The Hezron Mine, idle from 1912 to 1917, was reopened for a brief time during the war period before being shut down and dismantled in 1918. Kebler No. 1 (Tioga) and No. 2 (Big Four) were taken over in 1919. The Jobal Mine was purchased in 1920, at which time Rouse was abandoned after the flooding of underground workings.[7]

Operations in the Canon district remained relatively stable throughout the first two decades of the twentieth century. Nonac (Canon spelled backward), located immediately south of Canon City, opened in 1904 on land leased from the Santa Fe Railway. The small Emerald Mine near Williamsburg, initially operated on the Pueblo firm's property by Simon P. Smith, was purchased in 1919 to supply the wagon trade. The Brookside Mine closed in 1910 because excessive bone in the coal raised mining costs to prohibitive levels. However, lessees worked the deposit for several years to meet in part the local demand for fuel.[8]

All mines in the mountain district were adversely affected by the decline of the coke trade. The company suspended mining operations in Garfield and Pitkin Counties in 1916. In Gunnison County, the Anthracite Mine exhausted its available coal by the close of 1906. The Elk Mountain (Pershing) Mine, located a mile and a half north of Crested Butte, opened in 1918 as a replacement for the Floresta Mine, which closed the following year.[9]

A brief period of wartime prosperity was followed by devastating depression during the decade of the twenties, the consequence of excessive capacity and a shrinking market in which coal could not compete with petroleum, natural gas, and other forms of energy. In an effort to adjust production to demand while reducing the burden of overhead expenses, the fuel and iron company closed 13 properties in the years from 1922 through 1930. Victims of this retrenchment policy were Starkville, Engleville, Sopris, Berwind, Tabasco, and Primero in the Trinidad district; Jobal, Lester, and Kebler No. 1 near Walsenburg; and the Elk Mountain Mine in Gunnison County.[10]

Three more mines were abandoned during the darkest days of the Great Depression. Toller, Ideal, and Walsen closed as the company sought to adjust output to a contracting market. During the balance of the decade, the federal government's attempt to revitalize bituminous coal mining through the National Recovery Administration and the Guffey Acts temporarily stabilized the industry in Colorado. The Colorado Fuel and Iron Corporation's seven mines — Frederick and Morley near Trinidad ; Kebler No. 2, Pictou, and Robinson No. 4 (Cameron) at Walsenburg; Nonac in the Canon district; and Crested Butte — eventually enjoyed prosperity when mobilization for World War II required maximum production of all fuel resources.[11]

The Pueblo enterprise, in response to further changes in the fuel market, revived its policy of retrenchment after the close of World War II. The Robinson Mine ended production in 1946. Nonac and Crested Butte closed in 1952. Kebler No. 2 and Pictou shut down, in that order, in the following two years. Morley and Frederick ended operations in 1956 and 1960, respectively.[12] Meanwhile, the corporation developed in the Stonewall area west of Trinidad a large, automated mine to fill the fuel requirements of the steel works.

The Allen Mine, opened at Weston in 1951, yielded high-volatile coking coal. The underground workings featured continuous mining machines as well as conventional mobile cutting and loading units. Shuttle cars, conveyor belts, rope and rail haulage systems transported fuel to the surface where it was cleaned and run through breakers prior to shipment to Pueblo. There the product was washed, crushed, and blended with certain eastern coals prior to entering the by-product ovens.[13]

As the demand for coal declined in the twentieth century, the enterprise was forced to rely for income and profits more and more on metallurgical production. The reconstruction of the Minnequa Works, as part of Osgood's modernization program, had enlarged the facilities for making and fabricating iron and steel. By 1903, the plant was composed of five blast furnaces, two converters, six open-hearth furnaces, enlarged blooming and rail mills, plus units for the manufacture of rods, tinplate, hoops and ties, drawn wire, and nails.[14]

The remodeling and expansion scheme was interrupted twice in 1903, first by the company's threatened insolvency, and again by a strike of coal-mine and coke-oven workers in November, the effects of which lasted until October of the following year. Metallurgical production was halted part of that time in order to divert available fuel to commercial and domestic consumers.[15] The construction and refurbishing of production facilities was finally completed in 1907. By that date, a sixth blast furnace had been placed in operation, the number of open-hearth units doubled, and the efficiency of the metallurgical department improved with the construction of a calcining plant and a hot-metal storage tank. Additional soaking pits augmented the production of the rail mill. The unfinished second rod mill was redesigned to roll rods and squares for the enlarged bolt and spike mills. Also, a new conduit linked the Arkansas River and the water-storage basins south of the steel works.[16] These several projects marked the end of construction at the Pueblo site until just prior to World War I when work commenced on three more open-hearth furnaces and two batteries of by-product coke ovens.[17]

Another miners' strike, the most violent in Colorado's history, disrupted operations in 1913 and 1914. The Pueblo company insisted that the walkout occurred, not because of differences between employees and employer, but because the United Mine Workers of America, "an outside labor organization," sought to win recognition in Colorado. For a decade prior to 1913, company officials argued that "comparative harmony" had characterized the coal regions. Opportunities for employment were good, wages high, averaging four dollars or more daily, and mining laws were enforced. Indeed, the conditions "covered by the later demands of the labor organization, except those calling for recognition of the union and an increase of 10 per cent in wages, prevailed generally, if not entirely, over the State."[18] An overwhelming proportion of the firm's employees, it was said, resisted the union and had nothing to gain from joining its ranks.

Acting in the name of Colorado coal miners, union delegates meeting at Trinidad on September 15, 1913, issued several demands. In addition to recognizing the United Mine Workers of America as bargaining agent for employees, operators were to increase wages 10 percent for miners on tonnage, give an equal increase to coke-oven men, and adopt the daily wage scale that prevailed in Wyoming. The eight-hour day was to be observed in all mines and ovens. Coal diggers were to be given compensation for narrow and dead work, including brushing, timbering, removing rails, and handling impurities. Checkweighmen elected by miners were to be hired at all mines. Workers were to be granted the right to trade at any store, and to choose their own boarding houses and physicians. Mining laws were to be enforced, and the mine guard system was to be abolished.[19]

The Trinidad convention, the Pueblo company insisted, had not spoken for coal miners in southern Colorado. Delegates to that meeting were strikers from the northern lignite field, without employment for three and a half years, or men hand-picked by labor organizers rather than by locals whose interests they professed to serve. The vote in favor of a strike was not a surprise in view of the fact that the proceedings had been manipulated by officials of the United Mine Workers of America.[20]

The Colorado Fuel and Iron Company, as well as other operators in the southern field, rejected the union's demands, especially the claim to recognition as bargaining agent for coal miners. Only a few employees were union members, the Pueblo corporation stated, and a very large majority of the men did not wish to join the United Mine Workers of America. These experienced miners had come to Colorado because of the "open shop" and high wages, and they would not remain if required to submit to the domination of the labor organization.[21]

In Huerfano and Las Animas counties, where two-thirds of the state's and four-fifths of the Colorado Fuel and Iron Company's production was located, there was, the operator claimed, no appreciable support for the strike. When the walkout commenced on September 23, more than 60 percent of the Pueblo firm's employees remained on their jobs, and their numbers did not fall below that level in the next 7 months.[22]

Prior to the walkout, according to the fuel and iron corporation, the union had brought large quantities of arms and munitions into the southern counties. Reacting to the threat of from 1,500 to 2,000 fully armed strikers, the mining companies hastened to employ guards to protect their workmen and property.[23] However, experienced, efficient peace officers were unable to prevent anarchy in its worst form. The State Militia finally restored order. Men working for the operators voluntarily surrendered all arms; the strikers, however, turned in only a small proportion of their weapons.[24]

Peace was maintained until the militia was reduced to a handful of men. On the morning of April 20, 1914, a small detachment of soldiers stationed in the vicinity of Ludlow were assaulted by a force ten times their number. Lives were lost, including many women and children, the blame for which, according to the company, rested with the attackers. The incident marked the beginning of "a second reign of terror."[25] The United Mine Workers issued a call to arms; the Colorado Fuel and Iron Company reemployed armed guards and purchased rifles for workmen.[26] Peace returned to the region after ten days with the arrival of regular army troops, and the mines resumed operations at prestrike levels as the workers returned to their jobs.[27]

During and after the strike, the charge was frequently made that the coal companies controlled the militia. The Pueblo firm insisted that this was without foundation: relations with state troops "were the same in every particular as those that existed between us and the Federal soldiers."[28] Both were looked upon as peace officers, and their authority was respectfully recognized.

Numerous investigations have challenged the Colorado Fuel and Iron Company's view of the causes of the walkout and of the turbulent events of 1913-1914 in the southern counties. The United States Senate's Commission on Industrial Relations, after taking testimony from witnesses representing both labor and management, blamed the Pueblo corporation for conditions that brought about the strike and for most of the violence and bloodshed that characterized the seven-month dispute.[29] Historians, almost without exception, have refuted the argument that the United Mine Workers was the culprit, concluding that the company caused the unrest that led to the walkout, and that it was largely responsible for most of the violence that culminated in the so-called "Ludlow Massacre."[30]

Shocked by the violence that attended the strike, John D. Rockefeller, Jr., asked the Rockefeller Foundation to investigate ways of restoring harmony in the coal fields of Colorado. W. L. Mackenzie-King, author of Canada's labor code and later prime minister of that country, studied conditions on the western firm's properties, after which he proposed that employees exercise with management a voice in the determination of wages, conditions of labor, and other matters that directly affected the lives of workers.[31]

This proposal was an admission of sorts that the quality of life had deteriorated in the mill towns and mining camps, and that living conditions were in the eyes of company officials one of the causes of the miners' strike. Extant records suggest that the work of the Sociological Department had languished in recent years for lack of financial support.[32] Certainly, the testimony of several witnesses before the Commission on Industrial Relations revealed how far the Colorado Fuel and Iron Company had deviated from the goals set forth by Dr. Richard W. Corwin in launching the campaign for social betterment at the turn of the century.

Dr. Eugene S. Gaddis, a minister of the Methodist Episcopal Church and for a time superintendent of social work in all the Pueblo firm's mining camps, told the Commission on Industrial Relations that good dwellings were often in short supply, forcing miners and their families to reside in hovels, shacks, and dugouts that were unfit for habitation.[33] Water distributed at Lester could be used safely only for scrubbing floors, while foul-smelling seepage from a mine at Walsen was consumed in three communities. Educational facilities were inadequate in many camps and totally absent in some. The qualifications of teachers were known in a few instances to fall below minimum standards. Saloons, once banned from most camps and carefully regulated in others, flourished almost everywhere. Establishments for dispensing alcoholic beverages were encouraged by the company, which was notorious for its support of the liquor interests in southern Colorado. Although Gaddis' charges may have been exaggerated to some degree because of religious zeal, his views were largely substantiated by other witnesses who lived and worked in the company towns.[34]

The way to improve conditions for workers while at the same time avoiding a recurrence of strikes, Mackenzie-King suggested, was to reestablish personal relationships between labor and management. This was impossible in a literal sense because the company had 12,000 employees in 3 states. The practical solution was a system of representation by which employee-employer concerns could be discussed by spokesmen for each group. That scheme was implemented in the coal mines in 1915, and the following year in the Minnequa Works.[35]

The so-called "Rockefeller Plan" guaranteed all workers through a system of representation the right to bargain collectively, annual conferences with management, and numerous welfare benefits. The latter, with the exception of the long-established medical program, was administered by the Young Men's Christian Association.[36] This program of welfare capitalism probably lessened antagonisms for a time but did not eliminate, especially in the mines, periodic strikes. The company union, insofar as it applied to coal miners, remained in effect until 1933, when the corporation recognized the United Mine Workers of America as bargaining agent for employees in the fuel department.[37] Nine years later, the United Steelworkers of America became the spokesman for men in the Pueblo mills, at which time the Industrial Representation Plan was abandoned.[38]

World War I overshadowed the labor strife of 1913-1914. Even before the United States entered the conflict as a belligerent in 1917, the Colorado Fuel and Iron Company ran nearly at full capacity in all departments in an attempt to keep pace with the rising demand for metallurgical and fuel products. The Minnequa Works, about one-half of whose production was devoted to rails and accessories, was hard-pressed to fill all orders during the years from 1915 through 1919. Coal production, after lagging in recent years, soared briefly to record levels. Higher earnings permitted the company to pay arrears on preferred stock dividends and to make distributions for the first time since 1903 to common shareowners.[39]

Prosperity lasted only two years beyond the Armistice. A general business slump in the latter months of 1920 marked for the Pueblo enterprise the start of another period of prolonged financial stringency.[40] The metallurgical department operated for several years at less than two-thirds of capacity, with production devoted largely to rails and fastenings. At the same time, a steady decline in the fuel market prompted the curtailment of coal output and the closing of several mines.[41]

The Great Depression of the thirties further eroded the company's financial condition. The demand for railroad steel declined precipitously. Mounting deficits depleted working capital and undermined credit. The firm defaulted on its bond interest in 1933, leading to a receivership under the federal bankruptcy law. Following almost three years of legal proceedings, the United States District Court in Denver authorized a recapitalization plan which in effect turned the property over to the bond-holders. Since John D. Rockefeller, Jr., was the principal creditor, besides owning a large equity in the form of preferred and common shares, the reorganization did not diminish his influence in the successor enterprise, the Colorado Fuel and Iron Corporation.[42]

The reorganization plan, implemented on July 1, 1936, enabled the corporation to take advantage of improving conditions in the steel market. Net earnings for the first quarter of the fiscal year were two and one-half times greater than profits for the whole of the year 1935. Shipments for six months were higher than for any comparable period since 1930.[43] This trend, however, was short-lived. The recession of 1937 and 1938 delivered a staggering blow from which the company recovered only with extreme difficulty over a period of two years.[44]

While enjoying a brief period of relative prosperity in 1937, the Colorado Fuel and Iron Corporation diversified output and entered the Pacific Coast market for finished steel products by acquiring the California Wire Cloth Company. Purchased through an exchange of stock, the 58-year-old enterprise was a subsidiary of the Pacific Steel and Wire Company, with plants in Oakland and San Francisco.[45] These facilities had been for years consumers of Pueblo-made steel in the manufacture of a variety of products, including stucco netting, spring wire, fish trap netting, industrial screens, and wire cloth.[46]

World War II restored prosperity to heavy industry in the West. The fuel and iron corporation expanded output about 60 percent in producing throughout the period of national emergency rolled steel and ordnance for the armed forces. Construction of a new forging plant began in July 1940; the manufacture of heavy shells and casings commenced six months later. The Pueblo plant turned out in wartime nearly four million 155 mm. and 8-inch projectiles.[47]

Wartime alterations at Pueblo were designed to increase output of iron and steel for immediate military-oriented production and to strengthen at the same time the company's position in the postwar market. An ore bedding plant, the first of its kind in the United States, yielded a uniform ore mixture that enhanced the efficiency of the blast furnaces and reduced the volume of coke consumed in the production of pig iron.[48] Other wartime improvements included an additional blast furnace and the Minnequa Canal to transport water from the Arkansas River near Florence to reservoirs south of the steel works.[49]

Important changes were also instituted at the company's iron mines. At Sunrise, Wyoming, the only source of ore after the closing of the Orient and Fierro properties in the early years of the Great Depression, a shaft was excavated in 1942 to gain access to deposits from 500 to 700 feet below the surface. That year, the Pueblo firm also inaugurated mining operations near Cedar City in southwestern Utah. The Comstock Pit, the major producer in that area, yielded a mixture of magnetite and hematite. Thereafter, Wyoming and Utah ores were consumed in about equal proportions in the production of iron in Colorado.[50]

Wartime mobilization of the metallurgical industry ended the Colorado Fuel and Iron Corporation's uniqueness as the only fully integrated manufacturer of iron and steel products in the West. In 1942, the Kaiser Steel Company, with a loan from the Reconstruction Finance Corporation, erected blast furnaces, open-hearth units, and rolling mills at Fontana, California, 50 miles east of Los Angeles, to supply Henry J. Kaiser's shipbuilding activities. A similar plant was constructed by the Defense Plant Corporation at Geneva, Utah. Administered by the United States Steel Corporation, that facility turned out rolled steel for war production on the West Coast.[51]

At the war's end, the Colorado enterprise sought to take over and integrate with its Pueblo operations the government's Utah steel works. The fuel and iron corporation offered in 1945 to lease for five years the Geneva plant, with a renewal option, at a rental of $2.00 per ton on an estimated output of 500,000 tons of steel annually.[52] Simultaneously, Henry J. Kaiser bid for the Utah facility, and for a time he seriously considered the possibility of forming a western metallurgical syndicate to consolidate under one management steel production at Fontana, Geneva, and Pueblo.[53] The War Assets Administration finally sold the Utah property in 1946 to the United States Steel Corporation, but not before the Pueblo company had submitted an unsuccessful bid to buy rather than lease the plant.[54]

Expansion was the goal of Charles Allen, senior partner in the New York investment firm of Allen and Company, who with several associates gained control of the Colorado Fuel and Iron Corporation during the war years. In December 1944, they purchased the interests of John D. Rockefeller, Jr., and Rockefeller Center, Inc., amounting to $5,700,000 in bonds and 283,980 shares of common stock.[55] Because Allen was president of Wickwire Spencer Steel Company, speculation immediately centered on the possibility of a consolidation of that business venture with the Pueblo enterprise. A merger formally announced in mid-1945 enabled the western firm to further diversify its production to include a variety of light metal products while extending operations geographically into the eastern region of the United States.[56]

The Colorado Fuel and Iron Corporation secured through an exchange of stock Wickwire Spencer's iron and steel furnaces, rolling mills, and fabricating plants. The Pueblo firm added to its operations a basic steel plant and wire mills at Buffalo, New York; wire mills at Clinton, Palmer, and Worchester, Massachusetts; and a subsidiary, the American Wire Fabrics Corporation of Mt. Wolf, Pennsylvania, manufacturer of insect wire screening and industrial wire cloth.[57]

Additional properties were acquired in the East in the next several years. The Claymont (Delaware) Plant of the Worth Steel Company, purchased in 1951, added carbon and stainless plates, plus large-size gas and oil transmission pipes, to the product line.[58] In order to obtain ample supplies of pig iron for the Claymont mills, the Pueblo firm also bought the E. & G. Brooke Iron Company, operator of blast furnaces at Birdsboro, Pennsylvania, together with a subsidiary, the Richard Ore Company, owner of mines near Wharton, New Jersey.[59] Finally, the western enterprise secured in 1952 John A. Roebling and Sons Corporation, famous bridge contractor, with plants in Roebling and Trenton, New Jersey, devoted to the production of electrical wire and cable, bridge components, and wire rope.[60]

Production units were also modified or expanded at the Pueblo plant. The only manufacturer of steel rails and fastenings beyond the Mississippi River, the company moved away from almost complete dependence on those products. The wire mill was overhauled and modernized in anticipation of heavier loads on completion of a new rod mill in 1949, a replacement for the obsolete unit erected in 1903. Modern machinery turned out 25,000 tons of rods monthly compared to the previous yield of 10,000 for fabrication of wire, nails, fencing, netting, welded fabric, and barbed wire in a wide range of sizes and styles for the regional market.[61]

The long period of post-war prosperity ended for the Colorado Fuel and Iron Corporation in the closing years of the decade of the fifties. Like other American metallurgical enterprises, it had become increasingly vulnerable to competition from foreign manufacturers. Low-priced imports, especially wire products, undermined the domestic market with the result that many of the Pueblo firm's recent acquisitions could not be operated at a profit. Mounting deficits necessitated a major realignment. Plants were sold or closed.[62] By the end of 1971, after activities had been phased out at Palmer, production was restricted to Pueblo, South San Francisco, and Trenton-Roebling. The latter facilities, because of continuing losses, were shut down within less than two years.[63]

In the never-ending search for a greater share of the market, heightened by the intensity of competition from domestic as well as foreign steelmakers, the Colorado firm instituted improvements to increase output, expand product line, and reduce costs. A seamless tube mill completed in 1953 supplied the western oil industry with pipes and casings. A basic oxygen furnace, one of the first in the United States, began turning out steel in 1961, at which time operations of the open-hearth units were cut back.[64] The latter were in time superseded by another technological advance, an electric furnace that consumed only scrap metal in producing up to 300,000 tons of steel annually.[65] The last of the open hearths, dating from 1902, were deactivated after 70 years of service.

A five-year improvement program launched in 1968 further augmented production in western mills and mines. A continuous bar mill was erected at Pueblo to advance the output of coiled rods and squares, strips, reinforcing rods, and fence sections. One battery of the by-product coke ovens were replaced with modern units featuring air-quality controls. Funds were also allocated for the introduction of "longwall" mining at Weston and for development work in opening the Columbia Gulch iron-ore deposit near Sunrise, as well as for exploration of Utah iron deposits. Finally, structures and equipment were acquired to permit the adoption of "unit train systems" in moving coal from the Allen Mine to the steel plant.[66]

This improvement program was conceived and carried out by the CF&I Steel Corporation, a new name formally adopted on August 1, 1966. In abandoning "Colorado Fuel and Iron," the Pueblo firm recognized publicly what it had long known in private; the older name did not correctly identify the nature of the company's business activities because of changes that had occurred in recent decades. No longer was it a producer of coal or coke for the fuel trade. Iron was not an important item in its product line except as the basic metal for making steel. Furthermore, the enterprise was committed to a policy of developing new functions. It sought to enter into the manufacture of alloy steels and heavy machinery and components through acquisitions and mergers. To this end, the stockholders authorized in May 1967 the issuance of 1,000,000 shares of new preferred stock without declared par value.[67]

The additional equity capital was used to acquire subsidiaries in the Rocky Mountain region. The Denver-based Silver Engineering, Inc., renamed "CF&I Engineers" in 1969, was a metal-working and design-engineering firm that had been for many years a leader in the manufacture of beet and cane sugar harvesting and processing machinery. The company also specialized in custom designing and building equipment for food processing, mining, synthesis of chemicals, farming, and construction.[68] The Pueblo enterprise also purchased the assets of the Gilmore Steel Corporation of Salt Lake City. Renamed "CF&I Fabricators of Utah," that firm assembled reinforcing steel structures for construction projects. A similar concern, CF&I Fabricators, Inc., was located in Denver.[69]

The Pueblo firm was itself taken over by a corporation seeking to diversify operations. The Crane Company of New York announced in July 1969 that it controlled 82 percent of CF&I Steel's common stock. With manufacturing establishments in the United States, Canada, Mexico, the United Kingdom, France, Spain, The Netherlands, and Australia, Crane manufactured valves, pumps, water, and waste treatment equipment used in fluid and pollution control, plumbing, and heating, as well as aerospace products. Crane's business activities were consistent with the western firm's goal of diversification, and the latter acknowledged that it had much to gain from participation in the holding company's international operations.[70]

The pioneer steelmaker in the West, the CF&I Steel Corporation, celebrated on January 11, 1972, its 100th anniversary. The small venture created by General William J. Palmer and his associates to assist the narrow gauge Denver and Rio Grande Railway in building from Pueblo to Canon City had become in a century the region's largest heavy industry and the nation's twelfth-ranking steel manufacturer, measured by volume of shipments.[71]

The pioneer steelmaker in the West, through its predecessor enterprises, has played important roles in promoting the settlement of Colorado. The original venture, the Central Colorado Improvement Company, enabled the Denver and Rio Grande to build westward from Pueblo, opening a fertile and mineral-rich valley for commercial development. The Southern Colorado Coal and Town Company played a similar role in Huerfano and Las Animas counties. John Osgood's Colorado Fuel Company, together with affiliates, and Jerome B. Wheeler's Grand River Coal and Coke Company advanced coal mining in Garfield, Gunnison, and Pitkin counties, and contributed to the growth of the smelting industry in the mountain district.

The Colorado Coal and Iron Company erected and operated an integrated plant for the production of iron and steel. That activity accelerated the tempo of mining, manufacturing, and transportation in Colorado and adjoining states and territories. The firm marketed coal and coke; the latter, in addition to fueling blast furnaces in the reduction of iron ore, was vital to the processing of the output of precious-metal mines throughout the Centennial State. The merger of the Pueblo enterprise with Osgood's business interests brought into being in 1892 the Colorado Fuel and Iron Company, the leading heavy industry in the western one-third of the United States, and the ascendant firm in the lucrative regional fuel trade.

The successor CF&I Steel Corporation's 100th anniversary marked the pouring of the 60,000,000th ton of steel at the Pueblo plant. That historic heat of July 1972, like the first of April 1882, was rolled into rails. The initial product had been made in a Bessemer converter; that of the centennial year was produced in a basic oxygen furnace.[72]

Unquestionably, the West's requirements of steel products could have been supplied by eastern mills. However, in the absence of competition, the cost may have been exorbitant if not prohibitive; the purchase of metallurgical goods in Pittsburgh or Chicago would have aggravated the adverse flow of payments that hampered growth in the underdeveloped Rocky Mountain region. The Colorado enterprise, although its output of basic metals was small, enabled the region to avoid complete dependence on outside producers. Furthermore, in serving the western market, the Pueblo firm opened natural resources, founded communities, attracted capital and labor, employed several thousand workers, and stimulated a wide range of economic activities. By the close of the nineteenth century, the company was Colorado's largest employer, its major consumer of raw materials, and its principal user of transportation services, all of which had a beneficial effect on the Centennial State's well-being and prosperity.

The Colorado firm's activities eventually expanded beyond the Rocky Mountains to the Atlantic and Pacific coasts, and its products changed in response to the demands of the national market. The company's plants served several major industries (agriculture, construction, general manufacturing, metal working, mining, petroleum, quarrying, and transportation) with 18 categories of goods: chain link fabric, chemicals and coke, cold rolled steel, galvanized wire strand, grader blades and cutting edges, merchant wire and hardware, mining products, seamless casing and tubing, prestressing wire and strand, rails and accessories, semifinished and hot-rolled carbon steel, welded wire fabric, wire, wire rope, and woven belts and screens.[73]

The furnaces and mills that stand like sentinels along Pueblo's southern skyline remain a monument to General William Jackson Palmer's dream of a heavy industry in the West. The coal mines that launched the Central Colorado Improvement Company's business activities are gone, as are those of the Southern Colorado Coal and Town, Colorado Fuel, Grand River Coal and Coke, Colorado Coal and Iron, and similar corporations. Some of the camps survive, mere shadows of their former prominence; others have disappeared along with the throngs of miners who once extracted "black diamonds" throughout the southern field and mountain district. Only steel remains a viable economic force, but the current energy crisis may in time revive the fuel trade.

For one hundred years, the Pueblo enterprise has provided an industrial base for opportunity, prosperity, and stability in Colorado. The CF&I Steel Corporation is now engaged in its second century of service to the Centennial State and the Rocky Mountain region.

Subsidiary Corporations

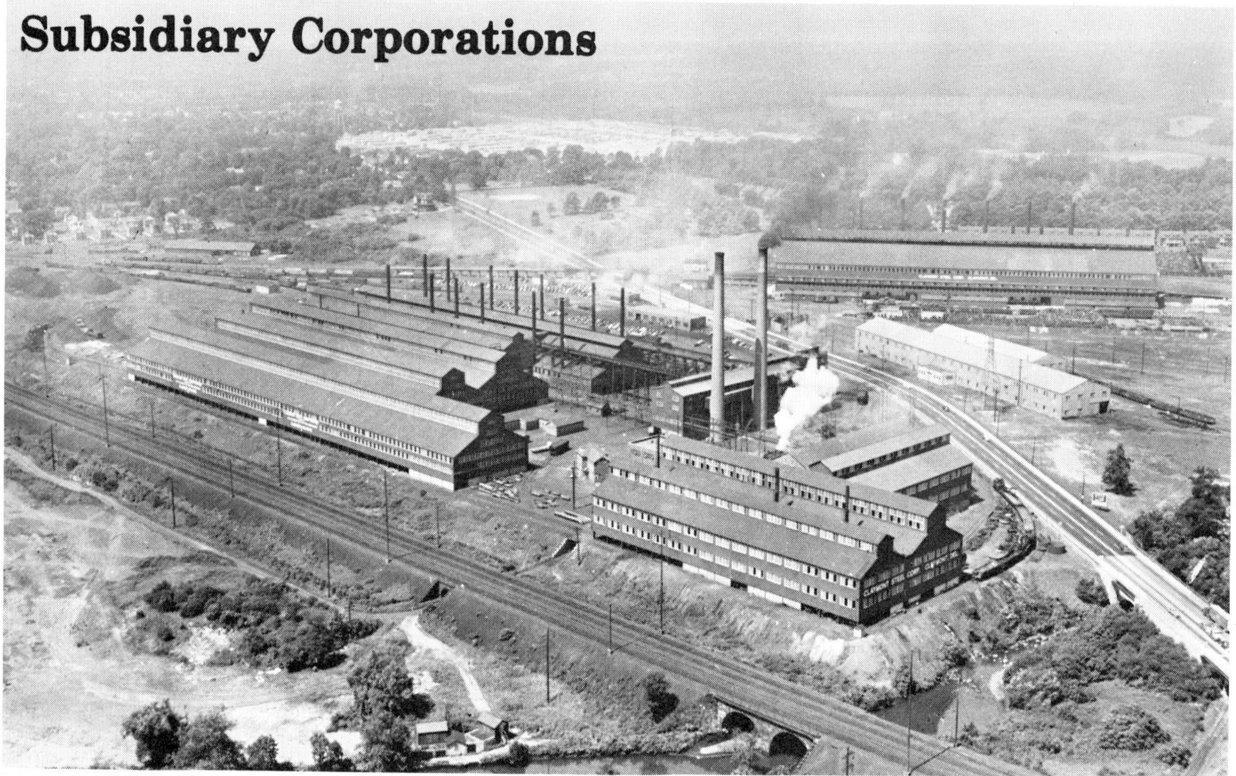

Plate and tube mills at the Clayton (Delaware) plant of the Worth Steel Co. This firm was acquired by the Colorado Fuel and Iron Corp. in 1951 as part of an on-going diversification program.

Aerial view of The California Wire Cloth Co.'s South San Francisco plant, 1946. The CF&I Steel Corporation purchased the 58-year-old California firm in 1937 in an attempt to diversify output and gain entrance into the Pacific Coast market for steel products.

John A. Roebling Sons Corporation's wire rope and cable plant, Roebling, N.J., 1968. The Colorado Fuel and Iron Corp. purchased Roebling's manufacturing facilities in 1952.

John A. Roebling Sons Corporation's wire mills, Trenton, N.J., 1966.

Wickwire Spencer Steel Co.'s woven wire products plant, Clinton, Mass., 1950.

Wickwire Spencer Steel Co.'s wire products plant, Palmer, Mass., 1950.

Blast furnaces of the Wickwire Spencer Steel Co., Buffalo, N.Y., purchased by the Colorado Fuel and Iron Corp. in 1947 as part of an expansion program that transformed a regional firm into a national manufacturer of steel products.

Wickwire Spencer Steel Co.'s dock, Buffalo, N.Y., 1950. This facility, along with other Wickwire Spencer properties, was acquired by the Colorado Fuel and Iron Corp. in 1947, as part of an expansion program carried out by Charles Allen and associates.

E.G. Brooke Iron Co.'s blast furnaces at Birdsboro, Pa., 1951. This acquisition was another step in the Pueblo firm's goal of product and market diversification.

Endnotes

Chapter 1. Introduction

[1] The origin and development of the early American iron industry is discussed at length in James M. Swank, *History of the Manufacture of Iron in All Ages, and Particularly in the United States from Colonial Times to 1891,* 2nd ed., (Philadelphia, 1892; New York, 1969), chaps. 10-25. See also Arthur Cecil Bining, *The Pennsylvania Iron Manufacture in the Eighteenth Century* (Harrisburg, 1938), *passim;* Victor S. Clark, *History of Manufactures in the United States* (3 vols., New York, 1929), I: 169-174.

[2] Clark, *History of Manufactures in the United States,* I: 499, 503, 507.

[3] Traditionally, railroads have been viewed as the principal stimulator of the iron industry and related enterprises in the United States in the two decades prior to the Civil War. This idea has been challenged by Robert W. Fogel who suggests that quantitatively the demand for rails did not dominate the development of the iron industry from 1840 to 1860. He denies the accuracy of the so-called "gross theory" but does concede that a small addition to iron consumption directly attributable to rails may have had far-reaching consequences on the evolution of the industry. See *Railroads and American Economic Growth: Essays in Econometric History* (Baltimore, 1964), chap. 5.

[4] Advances in furnace design and operation, including the substitution of mineral fuel for charcoal, are discussed in Peter Temin, *Iron and Steel in Nineteenth-Century America: An Economic Inquiry* (Cambridge, Mass., 1964), chaps. 2-4. See also Clark, *History of Manufactures in the United States,* I: 500-501.

[5] Temin, *Iron and Steel in Nineteenth-Century America,* pp. 44-47.

[6] *Ibid.*, p. 49.

[7] *Ibid.*, p. 48.

[8] *Ibid.*, pp. 49, 130-131.

[9] *Ibid.*, p. 50.

[10] Swank, *History of the Manufacture of Iron in All Ages,* pp. 396-397.

[11] Burton J. Hendrick, *The Life of Andrew Carnegie* (2 vols., Garden City, New York, 1932), I: 156-158.

[12] Swank, *History of the Manufacture of Iron in All Ages,* p. 400; Clark, *History of Manufactures in the United States,* II: 69-70, 257-258.

[13] Hendrick, *The Life of Andrew Carnegie,* I: 161-163.

[14] Swank, *History of the Manufacture of Iron in All Ages,* pp. 405-406.

[15] Clark, *History of Manufactures in the United States,* II: 62-63, 194-195.

[16] Swank, *History of the Manufacture of Iron in All Ages,* pp. 412-413; Clark, *History of Manufactures in the United States,* II: 74.

[17] Swank, *History of the Manufacture of Iron in All Ages,* p. 413.

[18] *Ibid.*, pp. 410-411.

[19] *Ibid.*, p. 413.

[20] James M. Swank, "Twenty Years of Progress in the Manufacture of Iron and Steel in the United States," American Iron and Steel Association, *Annual Statistical Report, 1892* (Philadelphia, 1893), p. 21.

[21] *Engineering and Mining Journal,* LXXI (Jan. 5, 1901), p. 5. A different price schedule for rails appears in American Iron and Steel Association, *Annual Statistical Report, 1900* (Philadelphia, 1901), p. 27.

[22] Swank, "Twenty Years of Progress in the Manufacture of Iron and Steel in the United States," pp. 7, 15.

[23] Temin, *Iron and Steel in Nineteenth-Century America,* pp. 179, 183.

[24] Excellent surveys of the gold rush and settlement of Colorado are in Jerome C. Smiley, ed., *Semi-Centennial History of the State of Colorado* (2 vols., Chicago, 1913), I: chaps. 9-22; James H. Baker and LeRoy R. Hafen, eds., *History of Colorado* (5 vols., Denver, 1927), II: chap. 8; LeRoy R. Hafen, ed., *Colorado and Its People: A Narrative and Topical History of the Centennial State* (4 vols., New York, 1948), I: chaps. 10-17; Carl W. Ubbelohde, *A Colorado History* (Boulder, Colo., 1965), chaps. 6-10.

[25] The expansion of mining activities in the decade of the seventies is treated in Hafen, ed., *Colorado and Its People,* I: 334-340; Ubbelohde, *A Colorado History,* pp. 155-165.

[26] Baker and Hafen, eds., *History of Colorado,* II: 457.

[27] The construction of railroads and their economic impact on Colorado is discussed in Hafen, ed., *Colorado and Its People,* I: chap. 23, II: chap. 24; Ubbelohde, *A Colorado History,* chap. 11. See also George L. Anderson, *General William J. Palmer: A Decade of Colorado Railroad Building, 1870-1880* (Colorado Springs, 1936), and *Kansas West* (San Marino, Calif., 1963); Robert G. Athearn, *Rebel of the Rockies: A History of the Denver and Rio Grande Western Railroad* (New Haven, 1962).

[28] Denver's rise as a smelting center began in 1878 when the Boston and Colorado Smelting Company opened its Argo Smelter on the outskirts of the city. The first important smelter was established in Pueblo in 1882 by Anton Eilers. See Baker and Hafen, (eds.), *History of Colorado,* II: 695-707; Hafen, ed., *Colorado and Its People,* I: 447-449; James Edward Fell, Jr., "The Boston and Colorado Smelting Company: A Study in Western Industrialism," (unpub. M.A. thesis, Department of History, University of Colorado, 1972), *passim.*

[29] The early metallurgical industry is discussed in Baker and Hafen, eds., *History of Colorado,* II: 712-719; Hafen, ed., *Colorado and Its People,* I: 454-455, II: 580-587. See also Albert B. Sanford, "The First Iron Manufactory in Colorado," *Colorado Magazine,* VIII (July, 1931): 142-146.

[30] The business activities of the Denver and Rio Grande Railway's ancillary enterprises are treated in Lois Marguerite Gaynor, "History of the Colorado Fuel and Iron Company and Constituent Companies, 1872-1933," (unpub. M.A. thesis, Department of Economics, University of Colorado, 1936), pp. 1-11.

[31] Colorado Coal and Iron Company, *Annual Report, 1879* (Colorado Springs, Colo., 1880) p. 9. The title of this publication is misleading since the coal and iron company was not incorporated until January 23, 1880. In fact, this is the sixth annual report of the Central Colorado Improvement Company and the second annual report of the Southern Colorado Coal and Town Company, the two of which were consolidated with the Colorado Coal and Steel Works Company on Dec. 13, 1879, to form the Colorado Coal and Iron Company.

[32] See agreement between the Denver and Rio Grande Railway Company and the Colorado Coal and Iron Company, Nov. 1, 1880, in D&RG Archives, The State Historical Society of Colorado, Denver.

[33] Colorado Coal and Iron Company, *Annual Report, 1881* (Colorado Springs, Colo., 1882), p. 13; *Annual Report, 1882* (Colorado Springs, Colo., 1883), pp. 11-13.

[34] Clark, *History of Manufactures in the United States,* II: 244; David Brody, *Steelworkers in America: The Nonunion Era* (Cambridge, Mass., 1960), p. 3.

35 U.S. Commissioner of Corporations, *Report on the Steel Industry* (Washington, D.C., 1911), pp. 2-13; Gertrude G. Schroeder, *The Growth of Major Steel Companies, 1900-1950* (Baltimore, 1953), pp. 36-46; Clark, *History of Manufactures in the United States,* II: 54-58.

Chapter 2. The Central Colorado Improvement Company

1 The early history of the Kansas Pacific and Denver Pacific railroads may be found in Herbert O. Brayer, "History of Colorado Railroads," in *Colorado and Its People,* LeRoy R. Hafen, ed., II: 635-647, and Anderson, *Kansas West, passim.*

2 John S. Fisher, *A Builder of the West: The Life of William Jackson Palmer* (Caldwell, Idaho, 1939), pp. 15-67.

3 *Ibid.*, pp. 68-125.

4 Anderson, *Kansas West,* pp. 63, 83.

5 Wm. J. Palmer, *Report of Surveys Across the Continent in 1867-'68, on the Thirty-fifth and Thirty-second Parallels, for a Route Extending the Kansas Pacific Railway to the Pacific Ocean at San Francisco and San Diego* (Philadelphia, 1869), *passim.* Palmer also proposed construction of a rail line from the Arkansas River over Raton Pass as an alternate route to Albuquerque.

6 Herbert O. Brayer, *William Blackmore: Early Financing of the Denver & Rio Grande and Ancillary Land Companies, 1871-1878* (Denver, 1949), pp. 28-40. See also Athearn, *Rebel of the Rockies,* p. 15.

7 Brayer, *William Blackmore: Early Financing of the Denver & Rio Grande,* pp. 28-40; Athearn, *Rebel of the Rockies,* p. 15.

8 Fisher, *A Builder of the West, passim.*

9 Brayer, *William Blackmore: Early Financing of the Denver & Rio Grande,* pp. 16, 18-19, 30-31.

10 *Ibid.*, p. 14.

11 Second ed., London and New York, 1870.

12 A biographical sketch of Bell is in "Death of Mr. W.A. Bell: A Loss to Blechingley and District," *The Surrey Mirror and County Post,* June 10, 17, 1921. A reprint of this article is in the William Abraham Bell Collection, The State Historical Society of Colorado, Denver. See also *National Cyclopedia of Biography* (51 vols., New York, 1898-1969), XXIV: 404.

13 Fisher, *A Builder of the West,* pp. 179, 194-196: Brayer, *William Blackmore: Early Financing of the Denver & Rio Grande,* pp. 19-20, 67n.

14 Fisher, *A Builder of the West,* pp. 179, 193-196; Brayer, *William Blackmore: Early Financing of the Denver & Rio Grande,* pp. 17, 21, 27, 67n.

15 Brayer, *William Blackmore: Early Financing of the Denver & Rio Grande,* p. 32.

16 *Ibid.*

17 *Ibid.*, pp. 49, 52.

18 *Ibid.*, p. 73.

19 *Ibid.*, pp. 33-34.

20 *Ibid.*, pp. 75-76. After only six months, the Mountain Base Investment Fund was succeeded by the National Land and Improvement Company, a joint-stock enterprise whose Pennsylvania charter empowered it to purchase, own, improve, and sell land in Utah, Arizona, and adjoining states and territories. The Mountain Base company had been limited to purchasing land.

21 *Ibid.*, pp. 80-83. See also Irving Howbert, *Memories of a Lifetime in the Pike's Peak Region* (New York, 1925), pp. 219-231.

22 Brayer, *William Blackmore: Early Financing of the Denver & Rio Grande,* pp. 83-84.

23 *Ibid.*, pp. 100-101.

[24] *Pueblo Daily Chieftain,* June 21, 1872.
[25] *Ibid.*, Jan. 11, Feb. 9, March 27, 28, Dec. 30, 1873.
[26] *Ibid.*, Nov. 20, Dec. 15, 31, 1872, Jan. 11, 1873.
[27] *Ibid.*, Jan. 1, 1873.
[28] Articles of association, November 11, 1871, and incorporation papers, Jan. 11, 1872, Film F. See also Central Colorado Improvement Company Minute Book, pp. 1-13.
[29] Subscription forms and list of subscribers, Film F.
[30] Brayer, *William Blackmore: Early Financing of the Denver & Rio Grande,* p. 147.
[31] Central Colorado Improvement Company Minute Book, pp. 37-40; "Agreement, Denver & Rio Grande Railway Company with Central Colorado Improvement Company for $1,040,000 Bonds ($825,000)," D&RG Archives.
[32] Central Colorado Improvement Company Minute Book, pp. 88-95.
[33] National Land and Improvement Company, *Annual Report, 1873* (Philadelphia, 1873), p. 18.
[34] *Ibid.*
[35] Brayer, *William Blackmore: Early Financing of the Denver & Rio Grande,* pp. 86-87.
[36] LeRoy Hafen, "Mexican Land Grants in Colorado," *Colorado Magazine,* IV (May, 1927): 86. See also U.S. House of Representatives, *Private Land Claims in New Mexico, House Ex. Doc.* 112, 37th Cong., 2 Sess. (Washington, 1862), pp. 36-38.
[37] "Statement, Brief, and Argument for the United States: Gervacio Nolan Grant, Private Claim No. 29," U.S. Court of Private Land Claims, *The Heirs of William Pinkerton vs. The United States et al.: Statement, Brief, and Argument for the United States* (Santa Fe, 1873), pp. 5-10.
[38] U.S. House of Representatives, *Report of the Committee on Private Land Claims, July 1, 1868. House Report* 71, 40th Cong., 2nd Sess. (Washington, 1868), p. 5.
[39] 16 U.S. *Statutes*, 646.
[40] Abstract No. 21880, File 46C-96, Land and Water Department, CF&I Steel Corporation, Pueblo.
[41] Henry Hodding to William Blackmore, Salisbury, March 20, 1871, and April 4, 1871, William Blackmore Papers, Records Center and Archives of New Mexico, Santa Fe.
[42] Bell to Blackmore, Denver, Dec. 18, 1871, *ibid.*
[43] *Pueblo Daily Chieftain,* March 21, 1875. See also correspondence between Henry C. Thatcher and Britton and Gray, Attorneys, Washington, D.C., Film F.
[44] Abstract No. 21880.
[45] Film M.
[46] Central Colorado Improvement Company, *Annual Report, 1874* (Colorado Springs, 1875), p. 10.
[47] *The Pueblo Colony of Southern Colorado* (Lancaster, Pa., 1874), p. 13; *Pueblo Daily Chieftain,* Nov. 1, 1872, Jan. 4, 5, 1873.
[48] Central Colorado Improvement Company, *Annual Report, 1877-1878* (Colorado Springs, Colo. 1879) pp. 9-10.
[49] *The Pueblo Colony of Southern Colorado,* p. 25.
[50] *Ibid.*; Central Colorado Improvement Company, *Annual Report, 1874,* p. 11.
[51] *The Pueblo Colony of Southern Colorado,* p. 26.
[52] *Pueblo Daily Chieftain,* Dec. 28, 1873, July 13, Aug. 27, Sept. 5, 1874. For a discussion of the difference between colony and town development schemes see James F. Willard, ed., *The Union Colony of Greeley, Colorado, 1869-1871* (Boulder, 1918), xviii-xix.

[53] The Colorado Fuel and Iron Company, "Predecessor Companies of The Colorado Coal & Iron Co.: The Central Colorado Improvement Co., The Southern Colorado Coal & Town Co., The Colorado Coal & Steel Works Co., and The Colorado Coal & Iron Company, pp. 4-5. The typescript of this statistical study of the predecessor companies is in the CF&I Steel Corporation's Comptroller's Office, Pueblo.

[54] R. C. Hills, "Colorado Coal Fields" (1886), p. 24. See also Hills, "Coal Lands of The Colorado Coal & Iron Co. in Colorado" (1892), pp. 99-100. Typescripts of Hills' reports on coal and iron resources are in the Mining Department, Pueblo.

[55] The U.S. Attorney for Colorado charged that Slocum and Musser had committed fraud in acquiring known mineral lands from the public domain as agricultural tracts. See J. A. Williamson to Carl Schurz, Washington, D.C., Feb. 8, 1879, in Department of the Interior Letters Received, National Archives, Washington, D.C.

[56] Film M.

[57] The Denver and Rio Grande made the decision to build to the coal field in November 1871 and did not test Canon coal in its locomotives until December of the same year. *Pueblo Daily Chieftain*, Nov. 23, Dec. 14, 1871.

[58] *Canon City Times*, Sept. 5, Oct. 17, 1872.

[59] *Pueblo Daily Chieftain*, Nov. 26, 1874.

[60] *Ibid.*

[61] National Land and Improvement Company, *Annual Report, 1873*, p. 17.

[62] Brayer, *William Blackmore: Early Financing of the Denver & Rio Grande*, p. 117.

[63] *Ibid.*, p. 153.

[64] *Pueblo Daily Chieftain*, Aug. 10, 1872.

[65] Film M.

[66] R. N. Clark to W. S. Jackson, Labran, June 11, 1874, in Denver and Rio Grande Railway Company, *Annual Report, 1874* (Colorado Springs, Colo., 1874), pp. 68-73.

[67] Film M. See also Central Colorado Improvement Company Minute Book, p. 124.

[68] Central Colorado Improvement Company, *Annual Report, 1874*, p. 14; *Annual Report, 1875-1876* (Colorado Springs, Colo., 1876), p. 8. Until 1880, the slope on Coal Creek was known as the Canon Mine. That year it was renamed the Coal Creek Mine. See Colorado Fuel and Iron Company, "Predecessor Companies of The Colorado Coal & Iron Co.," p. 3.

[69] Central Colorado Improvement Company, *Annual Report, 1877-1878*, p. 8.

[70] *Canon City Times*, Nov. 6, 1873; *Pueblo Daily Chieftain*, Nov. 22, Dec. 6, 1873, Sept. 25, 1875, Feb. 20, 1876, March 6, 1880.

[71] *Pueblo Daily Chieftain*, March 16, 1880.

[72] *Ibid.*, Jan. 30, 1878.

[73] Central Colorado Improvement Company, *Annual Report, 1877-1878*, p. 8.

[74] Colorado Fuel and Iron Company, "Predecessor Companies of The Colorado Coal & Iron Co.," p. 3.

[75] *Ibid.*, pp. 4-5.

[76] Two extant analyses of Grape Creek iron ore indicate a high metallic content but differ on the levels of impurities. See Robert H. Lamborn, "Report of Analysis by the Laboratory of the Pennsylvania Steel Co., June 19, 1871," and Andrew S. McGreath to Lamborn, July 15, 1872, in Denver and Rio Grande Scrapbook, Tutt Library, Colorado College, Colorado Springs.

[77] R. Lamborn to Mellen, Philadelphia, Feb. 17, 1872, Film F.

[78] Thomas M. Down to R. Lamborn, Philadelphia, March 4, 1872, in Denver and Rio Grande Railway Company, *Annual Report, 1874*, p. 89.

[79] Hunt to Palmer, Denver, Feb. 16, 1872, Film F.
[80] Mellen to Archer, Colorado Springs, Feb. 27, 1972, *ibid.*
[81] Archer to Mellen, Reading, Pa., March 16, 1872, *ibid.*
[82] R. Neilson Clark, "Pueblo. A Site for Smelting Works, Rolling Mills and Foundries," *The* (Pueblo) *People*, Feb. 21, 1874, clipping in The Central Colorado Improvement Company Scrapbook, Tutt Library, Colorado College.
[83] Central Colorado Improvement Company, *Annual Report, 1875-1876*, p. 9.
[84] *Canon City Times*, July 10, 1873.
[85] Denver and Rio Grande Railway Company, *Annual Report, 1874*, pp. 88, 93; Central Colorado Improvement Company, *Annual Report, 1874*, p. 19.
[86] *Pueblo Daily Chieftain*, Jan. 25, 1876; *Canon City Times*, Jan. 27, 1876.
[87] Lease dated Jan. 23, 1877, Film M; *Canon City Times*, March 29, 1877.
[88] Financial data for the Hot Springs appears in Colorado Fuel and Iron Company, "Predecessor Companies of The Colorado Coal & Iron Co.," p. 5, and Glenn W. Traer to John C. Osgood, Chicago, May 25, 1892.
[89] Terms of the trust deed of May 1, 1872, appear in Central Colorado Improvement Company Minute Book, pp. 32-37. See also *Annual Report, 1874*, p. 3.
[90] National Land and Improvement Company, *Annual Report, 1873*, pp. 19-20.
[91] Central Colorado Improvement Company, *Annual Report, 1874*, p. 4.
[92] *Pueblo Daily Chieftain*, Nov. 18, 1873.
[93] Central Colorado Improvement Company, *Annual Report, 1874*, pp. 5, 9.
[94] *Ibid.*, 1877-1878, p. 4.
[95] Colorado Fuel and Iron Company, "Predecessor Companies of The Colorado Coal & Iron Co.," p. 4.
[96] *Ibid.*
[97] "Memorandum of Agreement," July 15, 1878, appended to Central Colorado Improvement Company, *Annual Report, 1877-1878.* The earlier agreement of 1872 is also reprinted in this report.
[98] *Ibid.*, pp. 5, 9.
[99] "Proposition of Consolidation," in Central Colorado Improvement Company Minute Book, pp. 182-185.
[100] See Colorado Fuel and Iron Company, "Predecessor Companies of The Colorado Coal & Iron Co.," pp. 5-6.

Chapter 3. The Southern Colorado Coal and Town Company

[1] Certificate of incorporation on Film F. See also Colorado State Archives, Corporation Records.
[2] Brayer, *William Blackmore*, p. 156.
[3] *Ibid.*, p. 157; Athearn, *Rebel of the Rockies*, p. 61.
[4] Brayer, *William Blackmore*, p. 171.
[5] Palmer's negotiations in Europe are treated in *ibid.*, pp. 173-178.
[6] Agreement between the Denver and Rio Grande and W. A. Bell, June 30, 1875, and agreement between Bell and associates and W. S. Jackson, Trustee, June 30, 1875, in D&RG Archives.
[7] Agreement between Denver and Rio Grande and Union Contract Company, June 30, 1875, *ibid.*
[8] Brayer, *William Blackmore*, pp.
[9] Subscription paper in William A. Bell Collection. Arguments in support of the extension are presented in detail in Denver and Rio Grande Railway Company, *Annual Report, 1874*, pp. 62-77.
[10] The Cucharas, later called Walsenburg, coal was not cokeable, but it was an excellent steam and domestic fuel. The promoters may have been misled on this point by James T. Gardner's *Report Upon the Southern Coal and Iron Fields of Colorado Territory* (Colorado Springs, Colo., 1875), pp. 11, 15-16.
[11] Brayer, *William Blackmore*, p. 188.

[12] Agreement between W. A. Bell and Ernest P. Stephenson, Trustee, May 15, 1876, D&RG Archives.

[13] Agreement between the Southern Colorado Coal and Town Company and W. S. Jackson, Trustee, May 20, 1876, *ibid.*

[14] Agreement between Denver and Rio Grande Railway Company and W. S. Jackson, Trustee, June 20, 1876, *ibid.*

[15] Testimony of W. S. Jackson in *U.S. vs. The Colorado Coal and Iron Company*, Docket 668, U.S. Circuit Court, District of Colorado, in Federal Records Center, Region 8, Denver. Initiated against the Southern Colorado Coal and Town Company, this suit was amended in 1882 to make the successor enterprise the defendant.

[16] Notebook No. 3 (1876-1877), pp. 48-51, Robert F. Weitbrec Collection, State Historical Society of Colorado, Denver. See also testimony of W. S. Jackson, *U.S. vs. The Colorado Coal and Iron Company*, Docket 668.

[17] The value of each share was based on $322,900 in the stock of the Southern Colorado Coal and Town Company rather than $339,400 because 480 of the 9,885.8 acres were held by Jackson, Hunt, and WWB [William W. Borst] and were not included in the pool. See Notebook No. 3 (1876-1877), p. 50, Weitbrec Collection.

[18] Some members of the pool are identified in *ibid.*, pp. 48-49. See also testimony of W. S. Jackson in *U.S. vs. The Colorado Coal and Iron Company*, Docket 668.

[19] Southern Colorado Coal and Town Company, *Annual Report, 1878* (Colorado Springs, Colo., 1879), p. 3.

[20] *Pueblo Daily Chieftain*, Sept. 12, 1874.

[21] *Ibid.*, May 16, 1876.

[22] *Ibid.*, June 11, 1876.

[23] *Ibid.*, Jan. 23, 26, Feb. 9, March 18, June 10, 1876. The El Moro townsite was not officially platted until May 1877; see Colorado Fuel and Iron Company, "Predecessor Companies of The Colorado Coal & Iron Co.," p. 9.

[24] *Pueblo Daily Chieftain*, Jan. 23, 1876.

[25] Gardner, *Report Upon the Southern Coal and Iron Fields of Colorado Territory*, p. 8.

[26] "El Moro," *Camp and Plant*, I (May 17, 1902), p. 409. For the commercial rivalry between El Moro and La Junta, see *Pueblo Daily Chieftain*, June 10, 21, 25, 1876.

[27] *Pueblo Daily Chieftain*, June 25, 1876. See also Colorado Fuel and Iron Company, "Predecessor Companies of The Colorado Coal & Iron Co.," p. 9.

[28] Don Griswold and Jean Griswold, *Colorado's Century of "Cities"* (1958), pp. 137-138.

[29] *Ibid.*, p. 139; "Walsen and Robinson Mines," *Camp and Plant*, I (April 5, 1902): 265.

[30] Colorado Fuel and Iron Company, "Predecessor Companies of The Colorado Coal & Iron Co.," p. 8. See also *Pueblo Daily Chieftain*, March 29, 1876.

[31] *Pueblo Daily Chieftain*, March 29, May 6, 1876.

[32] Southern Colorado Coal and Town Company, *Annual Report, 1878*, p. 4.

[33] *Ibid.*; Colorado Coal and Iron Company, *Annual Report, 1879*, p. 6.

[34] "Camp Engle or Engleville," *Camp and Plant*, I (May 1, 1902): 457-458; Colorado Fuel and Iron Company, "Predecessor Companies of The Colorado Coal & Iron Co.," p. 151.

[35] Colorado Fuel and Iron Company, "Predecessor Companies of The Colorado Coal & Iron Co.," pp. 8-9.

[36] *Pueblo Daily Chieftain*, Dec. 20, 1878.

[37] See contract with Denver and Rio Grande Railway, May 20, 1876, in Southern Colorado Coal and Town Company, *Annual Report, 1878*, pp. 18-19.

[38] *Ibid.*, p. 5.

[39] There are no production records for the Walsen Mine during the time that it was worked under contract by Hunt and Allen.

[40] See production statistics in Colorado Fuel and Iron Company, "Predecessor Companies of The Colorado Coal & Iron Co.," p. 9.

[41] Southern Colorado Coal and Town Company, *Annual Report, 1878*, pp. 5-6. See also *The Blast*, May 13, 1932.

[42] Southern Colorado Coal and Town Company, *Annual Report, 1878*, p. 6; W. J. Palmer to Charles B. Lamborn, New York, Aug. 14, 1878, Film M.

[43] Frederick C. Steinhauer, "Coking of Coke," *Camp and Plant*, V (Jan. 23, 1904): 29-33.

[44] *Ibid.*

[45] Southern Colorado Coal and Town Company, *Annual Report, 1878*, pp. 9-10.

[46] *Ibid.*, p. 9; Colorado Coal and Iron Company, *Annual Report, 1879*, p. 7.

[47] Southern Colorado Coal and Town Company, *Annual Report, 1878*, p. 11; Colorado Fuel and Iron Company, "Predecessor Companies of The Colorado Coal & Iron Co.," p. 9.

[48] Agreement with Sebastian Stutz, March 3, 1879, Film M. See also Colorado Coal and Iron Company, *Annual Report, 1879*, p. 7.

[49] Colorado Coal and Iron Company, *Annual Report, 1879*, pp. 6-7.

[50] See bill of complaint, *U.S. vs. The Southern Colorado Coal and Town Company*, Docket 188, U.S. Circuit Court, District of Colorado, in Federal Records Center, Region 8, Denver.

[51] Hubbard to Drummond, Trinidad, Nov. 28, 1873. A certified copy, labeled Exhibit B, is attached to the testimony of witnesses in *U.S. vs. The Colorado Coal and Iron Company*, Docket 668.

[52] Drummond to Hubbard, Washington, Dec. 11, 1873, in *ibid.*

[53] *Pueblo Daily Chieftain*, Dec. 13, 1874.

[54] Taft to Decker, Washington, February 21, 1877, Department of Justice, Instruction Book G, p. 172; Decker to Charles Devens, Denver, Feb. 27, 1879, Department of Justice, Chronological File, Box 283, National Archives, Washington, D.C.

[55] Bill of complaint, *U.S. vs. The Southern Colorado Coal and Town Company*, Docket 188.

[56] Bill of complaint, *U.S. vs. The Southern Colorado Coal and Town Company*, Docket 647, U.S. Circuit Court, District of Colorado, in Federal Records Center, Region 8, Denver. This suit was held in abeyance while the federal government prosecuted Docket 668. After that case had been settled in favor of the defendant, on Jan. 31, 1888, Docket 647 was ordered discharged at cost of the plaintiff.

[57] *Trinidad Daily Advertiser*, April 14, 1888.

[58] J. A. Williamson to Carl Schurz, Washington, Feb. 8, 1879, Department of the Interior, Letters Received, National Archives.

[59] A description of lands acquired by the Trinidad Pool is in "Declaration of Trust," Jackson to Henry McAllister, Jr., Sept. 2, 1875, in D&RG Archives.

[60] Williamson to Schurz, Washington, Feb. 8, 1879, Department of the Interior, Letters Received.

[61] Decker to Devens, Denver, April 12, 1879, Department of Justice, Chronological File, Box 283.

[62] Schurz to Devens, Washington, May 2, 1879, Department of Justice, Chronological File, Box 93; Devens to Decker, Washington, May 6, 1879, Department of Justice, Instruction Book I.

[63] See records in *U.S. vs. The Colorado Coal and Iron Company*, Docket 668.

[64] See income account for the Southern Colorado Coal and Town Company in Colorado Fuel and Iron Company, "Predecessor Companies of The Colorado Coal & Iron Co.," pp. 9-10.

[65] Colorado Coal and Iron Company, *Annual Report, 1879*, p. 3.

Chapter 4. Colorado's First Rolling Mill

[1] *Pueblo Daily Chieftain*, Oct. 26, 1877.

[2] *Ibid.*, Aug. 29, 1877.

[3] *Ibid.*

[4] *Ibid.*, Sept. 6, 1877.

[5] *Ibid.*, Sept. 22, 1877.

[6] Draft of contract is in D&RG Archives.

[7] See articles of incorporation, Pueblo Iron Company, corporation records, Colorado State Archives.

[8] Subscription list, Pueblo Iron Company, Atchison, Topeka and Santa Fe Railroad Collection, RR Box 309, Kansas State Historical Society, Topeka. Cited hereafter as AT&SF Collection.

[9] List of expenses in relocating mill, Pueblo Iron Company, *ibid.*; *Pueblo Daily Chieftain*, Nov. 20, Dec. 5, 21, 1877, Jan. 1, 9, 1878.

[10] Subscription list, Pueblo Iron Company, AT&SF Collection, RR Box 309.

[11] *Pueblo Daily Chieftain*, Oct. 10, 1877.

[12] *Ibid.*, Jan. 27, 1878.

[13] Palmer to Faux, Colorado Springs, Feb. 5, 1878, in D&RG Archives. All transactions between the Pueblo Iron Co. and the Denver and Rio Grande Rwy. are in gross tons of 2,240 pounds.

[14] *Pueblo Daily Chieftain*, March 6, 1878. Other sources state operations from March 1. See Swank, *History of the Manufacture of Iron in All Ages*, p. 344, and Wilbur F. Stone, *History of Colorado* (4 vols., Chicago, 1918-1919), I: 547.

[15] Notebook No. 4 (1878), Weitbrec Collection.

[16] *Denver Daily Times*, Nov. 13, 1878. The process of making merchant iron, the initial step in rerolling rails, is described in J. B. Monahan, "Merchant Iron Department at the Minnequa Works in Pueblo," *Camp and Plant*, IV (July 25, 1903): 29-32.

[17] *Pueblo Daily Chieftain*, April 6, 1878.

[18] *Ibid.*, May 14, 1878.

[19] Notebook No. 4 (1878), Weitbrec Collection.

[20] *Denver Daily Times*, April 10, 1878. "Heavy iron" is the term for rails that weighed 30 or more pounds to the yard.

[21] *Ibid.*, May 27, 1878; *Rocky Mountain News*, May 29, 1878.

[22] M. C. Poor, *Denver South Park & Pacific: A History of the Denver South Park & Pacific and Allied Narrow Gauge Lines of the Colorado & Southern Railway Company* (Denver, Colorado, 1949), pp. 134, 144. See also *History of the City of Denver, Arapahoe County, and Colorado* (Chicago, 1880), pp. 589-591, 612-613.

[23] *Rocky Mountain News*, May 29, 1878.

[24] *Denver Daily Tribune*, May 30, 1878; *Denver Daily Times*, May 31, 1878.

[25] *Denver Daily Times*, June 3, 1878.

[26] *Pueblo Daily Chieftain*, June 18, 1878.

[27] *Ibid.* John Lambert was the proprietor and R. M. Stevenson was editor of the newspaper at this time.

[28] *Ibid.*, June 19, 1878.

[29] *Denver Daily Times*, June 18, 19, 1878.

[30] *Ibid.*, June 25, 1878.

[31] *Pueblo Daily Chieftain*, June 27, 1878.

[32] *Ibid.*
[33] *Ibid.*, July 4, 13, 26, 1878; *Denver Daily Times*, June 26, 1878.
[34] *Pueblo Daily Chieftain*, July 28, 1878.
[35] *Ibid.*, Aug. 3, 1878.
[36] *Ibid.*
[37] *Ibid.*
[38] *Ibid.*
[39] Articles of incorporation and certificate for corporate seal in Colorado State Archives, Corporation Records.
[40] *Denver Daily Tribune*, Sept. 14, 1878; *Pueblo Daily Chieftain*, Sept. 15, 1878.
[41] *Denver Daily Tribune*, Nov. 10, 1878.
[42] *Ibid.*, Nov. 13, 1878; *Pueblo Daily Chieftain*, Jan. 18, 1879.
[43] *Pueblo Daily Chieftain*, Dec. 21, 1878, Jan. 18, 1879.
[44] *Rocky Mountain News*, Feb. 21, 1880.
[45] *Ibid.*, April 25, July 27, 1879, Jan. 12, 1881.
[46] Production figures for Colorado are in James M. Swank, "Iron: Iron Ore and Its Products," in *Mineral Resources of the United States.* Edited by Albert Williams, Jr., *House Misc. Doc.*, No. 75, 48 Cong., 1 Sess. (1883), pp. 133, 136. The total output of Faux's mill is derived from Swank's data, minus the production for 1880, after it was purchased by the Colorado Coal and Iron Co. For the latter firm's production from the Denver Rolling Mill, see Colorado Fuel and Iron Company, "Predecessor Companies of the Colorado Coal & Iron Co.," p. 25.
[47] Evidence for this conclusion is drawn from Swank, "Iron: Iron Ore and Its Products," pp. 133, 136, and *Rocky Mountain News*, Feb. 21, 1880.
[48] Articles of incorporation, Film F. See also file for the Colorado Coal and Steel Works Company, corporation records, Colorado State Archives.
[49] Colorado Fuel and Iron Company, "Predecessor Companies of The Colorado Coal & Iron Co.," p. 11.
[50] *Ibid.*, pp. 6, 10-11.
[51] Colorado Coal and Iron Company, *Annual Report, 1880,* pp. 5, 10-11, Directors' Minute Book, I: 66.
[52] Colorado Coal and Iron Company, *Annual Report, 1884* (New York, 1885), p. 16.
[53] Colorado Fuel and Iron Company, "Predecessor Companies of The Colorado Coal & Iron Co.," p. 25.
[54] *Pueblo Daily Chieftain*, April 26, 1889.
[55] Following the demise of the Denver Rolling Mill, two attempts were made to establish in Denver an iron and steel plant to rival the facilities at Pueblo. The first began in 1889, when John W. Nesmith, president of the Colorado Iron Works, investigated at length the possibility of using coking coal and iron ore to produce pig iron in Gunnison County to supply steel manufacturing and fabrication mills in Denver. Nesmith also considered using iron ore from Hartsville, Wyoming. Although supported by prominent businessmen (for example, Henry M. Porter), Nesmith abandoned the scheme because the risks were inordinately high. See *Engineering and Mining Journal,* XLVII (March 23, 1889): 283, L (Sept. 27, 1890): 368, LVII (Feb. 3, 1894): 98; *Pueblo Daily Chieftain*, Jan. 30, 1902; Frank Hall, *History of the State of Colorado* (4 vols., Chicago, 1889-1895), IV: 153.

David H. Moffat and other Denver entrepreneurs organized in 1903 the Frontier Steel Co. to erect at Denver a plant to utilize coking coal and iron ore from northwestern Colorado, opened by the new Moffat Road. It was reported that summer that the Frontier Steel Co., the Northern Coal and Coke Company, and the Carbon Fuel Co. (the latter owned by the American Smelting and Refining Co.) had joined with Utah interest in forming the Utah Iron and Steel Co., which, under the direction of John C. Osgood, who had recently been ousted from the management of the Colorado Fuel and Iron Company, planned to compete with the Pueblo-based manufacturer of iron and steel products. Although the latter scheme failed to materialize, Moffat and his associates did entertain the hope of creating a metallurgical industry in Denver. See *Engineering and Mining Journal*, LXXV (Feb. 14, 1903): 271-272, LXXVI (July 11, 1903): 68, LXXVIII (July 28, 1904): 153; *Denver Post*, June 26, 27, 28, 29, 30, July 2, 1903.

Chapter 5. The Colorado Coal and Iron Company, 1880-1884.

1 Colorado Coal and Iron Company, *Annual Report, 1879*, p. 3.

2 Colorado Coal and Iron Company Minute Book, I:23. There are in the records for this enterprise three different books containing minutes of the meetings of directors and stockholders. For purposes of easy identification, they were well designated by number: I pertains to the period from Dec. 16, 1879, to April 7, 1884; II, from Dec. 16, 1879, to March 23, 1888; and III, from April 2, 1888, to Oct. 21, 1892.

3 Colorado Coal and Iron Company, *Annual Report, 1879*, p. 3.

4 *Ibid.*, pp. 44, 127. See also Colorado Fuel and Iron Company, "Predecessor Companies of The Colorado Coal & Iron Co.," pp. 41-42; *Pueblo Daily Chieftain*, Jan. 8, 1881.

5 Colorado Fuel and Iron Company, "Predecessor Companies of The Colorado Coal & Iron Co.," p. 43; *Pueblo Daily Chieftain*, Oct. 4, 1881.

6 Colorado Coal and Iron Company, *Annual Report, 1880*, p. 8.

7 Colorado Coal and Iron Company, *Annual Report, 1879*, p. 12.

8 *Ibid.*, p. 4.

9 *Ibid.*

10 Colorado Fuel and Iron Company, "Predecessor Companies of The Colorado Coal & Iron Co.," p. 23; Colorado State Inspector of Coal Mines, *Annual Report, 1884* (Denver, 1886), p. 10.

11 Colorado Coal and Iron Company, *Annual Report, 1879*, pp. 7-9.

12 Agreement in D&RG Archives. See also Clark, *History of Manufactures in the United States*, II: 531-532.

13 Colorado Coal and Iron Company, *Annual Report, 1879*, p. 9.

14 *Pueblo Daily Chieftain*, Sept. 8, 1881.

15 "Memories of Early Days at Minnequa," *Colorado Fuel & Iron Company Industrial Bulletin*, II (Oct. 31, 1916): 12-13.

16 Recollections of Bill Donahue in *CF&I Blast*, Feb. 4, 1938.

17 *Pueblo Daily Chieftain*, Feb. 10, 1881.

18 *Ibid.*, March 1, 12, Aug. 10, 1881.

19 Colorado Coal and Iron Company, *Annual Report, 1880*, p. 5, 10-11. Production statistics for the Denver Rolling Mill are in Colorado Fuel and Iron Company, "Predecessor Companies of The Colorado Coal & Iron Co.," p. 25.

20 Colorado Fuel and Iron Company, "Predecessor Companies of The Colorado Coal & Iron Co.," p. 19; *Pueblo Daily Chieftain*, April 26, June 2, 1889.

21 Colorado Coal and Iron Company, *Annual Report, 1881*, p. 14.

22 *Ibid.*; *Pueblo Daily Chieftain*, March 12, Oct. 14, 1881.

23 *Steel Works Blast*, Dec. 13, 1923, Jan. 12, 1928.

24 Colorado Coal and Iron Company Minute Book, I: 156-157; Danforth to Jones, Colorado Springs, April 6, 1881, Film M.

25 Colorado Coal and Iron Company, *Annual Report, 1881*, p. 13; Colorado Fuel and Iron Company, "Predecessor Companies of The Colorado Coal & Iron Co.," p. 17. See also *Pueblo Daily Chieftain*, Sept. 8, 1881.

26 *Pueblo Daily Chieftain*, March 12, April 2, 1882.

27 Colorado Coal and Iron Company, *Annual Report, 1880*, pp. 8-9; *Steel Works Blast*, Dec. 13, 1923.

[28] Colorado Coal and Iron Company, *Annual Report, 1881*, p. 13; *Pueblo Daily Chieftain*, Oct. 7, 1882, Nov. 18, 1883; Notebook No. 10 (1882), Weitbrec Collection.

[29] Colorado Coal and Iron Company, *Annual Report, 1881*, p. 15.

[30] *Pueblo Daily Chieftain*, March 9, July 8, 1902.

[31] *Ibid.*, March 9, 1902.

[32] *Ibid.*, April 30, 1882, March 9, 1902.

[33] *Ibid.*, April 30, June 18, 25, 29, 1882. Annual production statistics for the steel works are in Colorado Fuel and Iron Company, "Predecessor Companies of The Colorado Coal & Iron Co.," p. 25.

[34] *CF&I Blast*, Sept. 23, 1932.

[35] Colorado Coal and Iron Company, *Annual Report, 1881*, p. 16; *Pueblo Daily Chieftain*, April 2, 1882.

[36] Colorado Coal and Iron Company, *Annual Report, 1882*, p. 13. See also Colorado Fuel and Iron Company, "Predecessor Companies of The Colorado Coal & Iron Co.," p. 18.

[37] Colorado Coal and Iron Company, *Annual Report, 1882*, p. 13. The manufacture of cut nails is described in *Pueblo Daily Chieftain*, Aug. 6, 1882.

[38] *Steel Works Blast*, March 6, 1924; *CF&I Blast*, Aug. 2, 1929.

[39] *Pueblo Daily Chieftain*, Nov. 3, 5, Dec. 5, 7, 1882.

[40] Colorado Coal and Iron Company, *Annual Report, 1882*, p. 13; Colorado Fuel and Iron Company, "Predecessor Companies of The Colorado Coal & Iron Co.," p. 18.

[41] *Pueblo Daily Chieftain*, Oct. 10, Dec. 27, 1882.

[42] Either 922 or 823.21 tons of Grape Creek ore was shipped to Bessemer. The first figure appears in Colorado Coal and Iron Company, *Annual Report, 1882*, p. 14, and the latter in Colorado Fuel and Iron Company, "Predecessor Companies of The Colorado Coal & Iron Co.," p. 21.

[43] "Official Statements and Reports: The Colorado Coal and Iron Company," *Engineering and Mining Journal*, XXXVIII (July 26, 1884): 52. Cited hereafter as "Official Statements and Reports."

[44] Colorado Coal and Iron Company, *Annual Report, 1880*, pp. 5, 9-10; *Annual Report, 1881*, pp. 17-19. The lease for the Placer Mine, dated Sept. 25, 1880, is on Film M.

[45] For production of iron mines, see Colorado Fuel and Iron Company, "Predecessor Companies of The Colorado Coal & Iron Co.," p. 24.

[46] Contracts between the company and Salteel, dated March 4 and June 16, 1881, Film M; correspondence between Danforth and Haumann related to the Placer Mine, Film H. See also Rolker's report summarized in "Official Statements and Reports," pp. 52-55; Colorado Coal and Iron Company, *Annual Report, 1883* (New York, 1884), p. 11.

[47] Colorado Coal and Iron Company, *Annual Report, 1881*, p. 17.

[48] Haumann to Danforth, Colorado Springs, July 1, 1881, and Danforth to Palmer, Feb. 4, 1881.

[49] *The Blast*, March 13, 1931.

[50] *Ibid.*; "Orient Iron Mines," *Camp and Plant*, I (April 19, 1902): 314.

[51] Colorado Coal and Iron Company, *Annual Report, 1880*, p. 10. See also contracts with Breece Mining Company, Nov. 30, 1880 and Jan. 10, 1881, Film M.

[52] Colorado Coal and Iron Company, *Annual Report, 1883*, p. 10; *Pueblo Daily Chieftain*, Sept. 30, Oct. 4, Nov. 25, 1883. See also Clark, *History of Manufactures in the United States*, II: 257-258.

53 Colorado Coal and Iron Company, *Annual Report, 1883,* p. 16; "The Limestone Quarries at Lime," *Camp and Plant,* II (July 5, 1902): 1; *Pueblo Daily Chieftain,* Feb. 19, Sept. 24, Oct. 29, 1882, Sept. 12, 1886; *Steel Works Blast,* Dec. 13, 1923; Colorado Fuel and Iron Company, "The Predecessor Companies of The Colorado Coal & Iron Co.," p. 21.

54 Colorado Coal and Iron Company, *Annual Report, 1879,* p. 7; *Annual Report, 1880,* p. 7; *Annual Report, 1881,* p. 11; *Annual Report, 1882,* p. 10; *Annual Report, 1883,* p. 9. F. C. Miller, "History of Coal Development in Colorado, 1880-1900," *Colorado Fuel & Iron Company Industrial Bulletin,* XII (April, 1927): 26. See also Rolker's report in "Official Statements and Reports," p. 53.

55 Colorado Coal and Iron Company, *Annual Report, 1884,* p. 12.

56 See Rolker's report in "Official Statements and Reports," p. 53, for comparative costs of manufacturing iron and steel at Bessemer and in Pennsylvania.

57 The steel company's complaints against the railway are set forth in Danforth to Palmer, Nov. 3, 1882, D&RG Archives. See also Colorado Coal and Iron Company, *Annual Report, 1884,* p. 16.

58 Danforth to Palmer, Nov. 3, 1882, D&RG Archives.

59 Colorado Coal and Iron Company, *Annual Report, 1883,* pp. 6, 10, 13; *Pueblo Daily Chieftain,* Dec. 23, 1883.

60 Colorado Coal and Iron Company, *Annual Report, 1883,* pp. 6, 10, 13.

61 Colorado Coal and Iron Company Minute Book, I: 172-173.

62 *Ibid.,* I: 194-195, 198.

63 See records for *U.S. vs. The Southern Colorado Coal and Town Company,* Dockets 188 and 647, U.S. Circuit Court, District of Colorado, Federal Records Center, Region 8, Denver.

64 See records for Docket 668, U.S. Circuit Court, District of Colorado, Federal Records Center, Region 8, Denver. Initially this case was instituted against the Southern Colorado Coal and Town Company. On Jan. 7, 1882, the bill of complaint was amended making the Colorado Coal and Iron Company, Louis H. Myer, Frank D. Bloom, Mahlon D. Thatcher, and John A. Thatcher parties to the suit.

65 Colorado Coal and Iron Company, *Report of the General Counsel, April, 1884* (New York, 1884), pp. 10-11.

66 Judge McCrary's decision, *U.S. vs. The Colorado Coal and Iron Company,* Docket 668.

67 *Trinidad Daily Democrat,* Sept. 5, 1882.

68 Colorado Coal and Iron Company Minute Book, II: 139.

69 Colorado Coal and Iron Company, *Report of the General Counsel,* p. 10. See also records in *U.S. vs. The Colorado Coal and Iron Company,* Docket 1453, U.S. Circuit Court, District of Colorado, in Federal Records Center, Region 8, Denver.

70 Colorado Coal and Iron Company Minute Book, I: 200.

71 *Pueblo Daily Chieftain,* March 14, 1884.

72 Adolph Engler to Board of Directors, Colorado Coal and Iron Company, New York, March 3, 1884, in *Commercial and Financial Chronicle,* XXXVIII (March 8, 1884): 294; *Pueblo Daily Chieftain,* March 15, 1884.

73 C. W. Drake to A. Engler, New York, March 6, 1884, in *Commercial and Financial Chronicle,* XXXVIII (March 8, 1884): 295; *Pueblo Daily Chieftain,* March 15, 1884.

74 See Rolker's report in "Official Statements and Report," pp. 53-54; Clark, *History of Manufactures in the United States,* II: 531-532.

75 *Commercial and Financial Chronicle,* XXXVIII (March 8, 1884): 294.

76 *Ibid.,* March 1, 1884, p. 261.

[77] The text of the resolution by Curtis is in *Rocky Mountain News*, March 27, 1884.
[78] *Commercial and Financial Chronicle*, XXXVIII (March 29, 1884): 379.
[79] *Rocky Mountain News*, April 8, 1884.
[80] Colorado Coal and Iron Company Minute Book, II: 206-209.
[81] *Ibid.*, II: 204-205.
[82] *Ibid.*, II: 211.
[83] William S. Jackson, "General W.J. Palmer," p. 18. Typescript in The State Historical Society of Colorado.
[84] *Ibid.*, pp. 15-17.
[85] Colorado Coal and Iron Company, *Annual Report, 1882*, p. 22.

Chapter 6. The Colorado Coal and Iron Company, 1885-1892.

[1] Colorado Coal and Iron Company, *Annual Report, 1884*, p. 15; Rolker's report in "Official Statements and Reports," p. 53.
[2] Colorado Coal and Iron Company, *Annual Report, 1884*, p. 16.
[3] *Ibid.*, pp. 7-8, 16.
[4] *Ibid.*, p. 8.
[5] *Pueblo Daily Chieftain*, Nov. 24, 1872, Jan. 24, June 12, 1873, March 10, 28, 1875, June 6, 8, 1877.
[6] Colorado Coal and Iron Company, *Annual Report, 1879*, pp. 5-6; *Pueblo Daily Chieftain*, Dec. 4, 1881.
[7] *Pueblo Daily Chieftain*, April 26, 1882.
[8] See "An Act to Regulate the Working and Inspection of Coal Mines," in Colorado State Inspector of Coal Mines, *Annual Report, 1884*, pp. 137-146; U.S. Commissioner of Labor, *Report of Strikes and Lockouts, Jan. 1, 1881-Dec. 31, 1886, House Ex. Doc.* No. 1, Pt. 5, 50 Cong., 1 Sess. (5 vols., Washington, 1887), V: 52-55.
[9] Colorado Coal and Iron Company Minute Book, II: 212-213.
[10] *Pueblo Daily Chieftain*, Aug. 31, Nov. 25, 1884.
[11] *Ibid.*
[12] *Ibid.*, Aug. 29, 31, Nov. 25, 1884; (Denver) *Labor Enquirer*, April 26, June 9, July 26, 1884.
[13] *Pueblo Daily Chieftain*, Aug. 22, 1884.
[14] *Ibid.*, Aug. 29, 30, 1884; (Denver) *Labor Enquirer*, Sept. 13, 1884.
[15] *Trinidad Daily Advertiser*, Oct. 27, 29, 30, 1884.
[16] *Pueblo Daily Chieftain*, Nov. 1, 1884; (Denver) *Labor Enquirer*, Nov. 1, 1884.
[17] *Denver Tribune-Republican*, Nov. 4, 1884; (Denver) *Labor Enquirer*, Nov. 8, 1884.
[18] Colorado Coal and Iron Company, *Annual Report, 1884*, p. 9.
[19] *Pueblo Daily Chieftain*, Nov. 25, 27, 1884.
[20] *Ibid.*, Nov. 11, 13, 14, 19, 20, 25, 27, 1884.
[21] *Ibid.*, Nov. 26, 27, 1884: *Trinidad Daily Advertiser*, Nov. 21, 1884.
[22] *Pueblo Daily Chieftain*, Dec. 17, 1884; *Trinidad Daily Advertiser*, Dec. 19, 1884; contract between the Colorado Coal and Iron Company and miners at Coal Creek, Dec. 20, 1884, Film M.
[23] *Trinidad Daily Advertiser*, Jan. 15, 17, 1884; *Pueblo Daily Chieftain*, Jan. 20, Feb. 3, 1885. See also U.S. Commissioner of Labor, *Report of Strikes and Lockouts, Jan. 1, 1881-Dec. 31, 1886*, V: 52-55.
[24] *Trinidad Daily Advertiser*, Jan. 23, 25, 1885; *Pueblo Daily Chieftain*, Jan. 29, 1885.
[25] *Pueblo Daily Chieftain*, Feb. 17, 18, 22, 1885.
[26] *Ibid.*, March 12, 1885; (Denver) *Labor Enquirer*, Feb. 28, March 14, 1885.
[27] (Denver) *Labor Enquirer*, March 14, 1885.

28 Brief strikes occured at Coal Creek in 1889 and 1890. See *Pueblo Daily Chieftain*, Aug. 14, 1889, Jan. 15, 1890. A new labor contract, dated Aug. 16, 1889, is on Film K.

29 *Pueblo Daily Chieftain*, May 27, 1885.

30 *Ibid.*, May 28, 1885.

31 See records for *Edmund D. Davis vs. The Colorado Coal and Iron Company*, U.S. Circuit Court, District of Colorado, Docket 1993, in Federal Records Center, Region 8, Denver.

32 *Pueblo Daily Chieftain*, Jan. 24, 25, Feb. 1, 1885.

33 *Ibid.*, March 29, 31, 1885.

34 *Ibid.*, July 22, Sept. 27, 30, Nov. 8, 1885; Colorado Coal and Iron Company, *Annual Report, 1885* (New York, 1886), p. 13.

35 Colorado Coal and Iron Company, *Annual Report, 1885*, pp. 13, 21.

36 *Ibid.*, 1884, p. 16.

37 "Official Statements and Reports," p. 53.

38 Colorado Coal and Iron Company, *Annual Report, 1885*, p. 12.

39 Colorado Coal and Iron Company Minute Book, II: 263; *Pueblo Daily Chieftain*, Nov. 8, 1885.

40 Colorado Coal and Iron Company Minute Book, II: 270.

41 *Ibid.*, pp. 286, 288, 298-299.

42 Colorado Coal and Iron Company, *Annual Report, 1886* (New York, 1887), p. 10; *Pueblo Daily Chieftain*, July 18, 1886.

43 Contract dated Aug. 1, 1886, Film O. See also *Pueblo Daily Chieftain*, Sept. 3, 1886.

44 *Pueblo Daily Chieftain*, Aug. 16, 22, 1886.

45 *Ibid.*, Sept. 1, 1886.

46 *Ibid.*, Sept. 17, Nov. 21, 28, Dec. 22, 1886.

47 *Ibid.*, Sept. 17, Nov. 24, 28, 1886.

48 Colorado Coal and Iron Company, *Annual Report, 1886*, p. 11.

49 *Pueblo Daily Chieftain*, Nov. 7, 1886. See also Colorado Coal and Iron Company, *Annual Report, 1887* (New York, 1888), p. 11.

50 Contract, dated Oct. 21, 1886, Film O; *Pueblo Daily Chieftain*, Nov. 14, 18, Dec. 15, 1886.

51 *Pueblo Daily Chieftain*, April 3, Oct. 26, 29, Dec. 31, 1887.

52 *Ibid.*, June 11, 1887.

53 Colorado Coal and Iron Company, *Annual Report, 1886*, p. 9. See also agreement between the coal and iron company, the Grand River Coal and Coke Company, the Colorado Midland Railway, and the Aspen and Western Railway, dated March 18, 1887, Film O.

54 Colorado Coal and Iron Company, *Annual Report, 1887*, p. 10; Aspen and Western Railway Company Minute Book, pp. 22-50.

55 Colorado Fuel and Iron Company, "Predecessor Companies of The Colorado Coal & Iron Co.," p. 20.

56 Colorado Coal and Iron Company, *Annual Report, 1887*, p. 11; *Engineering and Mining Journal*, XLIII (April 9, 1887): 262.

57 Colorado Coal and Iron Company, *Annual Report, 1888* (1889), p. 13.

58 Clark, *History of Manufactures in the United States*, II: 334; *Engineering and Mining Journal*, XLV (June 20, 1888): 478.

59 Colorado Coal and Iron Company, Minute Book, III: 30.

60 *Pueblo Daily Chieftain*, Oct. 21, Dec. 15, 1888.

61 Colorado Coal and Iron Company Minute Book, III: 42-43, 50-51.

62 *Pueblo Daily Chieftain*, March 10, 1889, Aug. 16, 1901; contract between the Colorado Fuel and Iron Company and the Florence Oil and Refining Company, dated Dec. 12, 1892, Film K.

[63] Colorado Coal and Iron Company Minute Book, III: 31, 39.
[64] *Ibid.*, p. 59; *Pueblo Daily Chieftain*, Dec. 11, 1889.
[65] Colorado Coal and Iron Company Minute Book, III: 123, 134; DuPuy to Dougherty, Pueblo, April 15, 1889, Film H; *Pueblo Daily Chieftain*, April 2, 21, 1889, June 3, 1890.
[66] *Pueblo Daily Chieftain*, April 2, 1889.
[67] *Ibid.*, Sept. 4, Oct. 1, 1889.
[68] *Ibid.*, Oct. 22, 1889, April 20, 1890.
[69] *Ibid.*, June 2, 16, 1889.
[70] *Ibid.*, April 27, 1890.
[71] *Ibid.*, July 27, 1890, Nov. 13, 1892.
[72] DuPuy to President and Board of Directors, Pueblo, Jan. 29, 1889.
[73] Contract between the coal and iron company and Ramsay, dated May 1, 1889, Film I: *Rocky Mountain News*, March 25, 1892.
[74] Colorado Coal and Iron Company Minute Book, III: 173; *Pueblo Daily Chieftain*, Nov. 22, Dec. 15, 1889.
[75] *Pueblo Daily Chieftain*, Nov. 9, 26, 1889. Furnace No. 2 was named in honor of Mary Steck, wife of the general manager.
[76] *Ibid.*, Jan. 13, June 9, July 7, 16, Nov. 3, 1889.
[77] *Ibid.*, March 9, 30, April 27, May 11, 25, June 5, 15, 22, 29, Aug. 3, 17, Sept. 28, Oct. 26, Nov. 16, Dec. 3, 1890; Colorado Coal and Iron Company, *Annual Report, 1890* (1891), p. 11.
[78] *Pueblo Daily Chieftain*, Feb. 16, March 2, 1890.
[79] *Pueblo Daily Chieftain*, Feb. 2, 9, 16, 23, March 26, April 20, 1890.
[80] *Ibid.*, March 13, 1890.
[81] *Ibid.*, April 13, 20, 1890. In its rebuilt condition, Furnace No. 1 was christened "Lillie-Georgia" in honor of the daughter and wife, respectively, of general manager R. G. Ford.
[82] *Ibid.*, April 13, 20, 27, 30, May 4, 11, June 22, Aug. 17, 30, 1890; Colorado Coal and Iron Company, *Annual Report, 1890*, p. 10.
[83] *Pueblo Daily Chieftain*, May 25, Sept. 12, 1890; Colorado Coal and Iron Company, *Annual Report, 1890*, p. 10.
[84] Colorado Coal and Iron Company, *Annual Report, 1890*, p. 11.
[85] *Ibid., 1891* (1892), pp. 8-9.
[86] *Pueblo Daily Chieftain*, Jan. 11, 1891.
[87] *Ibid.*, May 13, June 7, 1891.
[88] Colorado Coal and Iron Company, *Annual Report, 1891*, p. 9.
[89] *Pueblo Daily Chieftain*, Aug. 23, Sept. 10, Oct. 25, Nov. 15, Dec. 27, 1891.
[90] Colorado Fuel and Iron Company, "Predecessor Companies of The Colorado Coal & Iron Co.," p. 32. Extant records do not indicate the nature of the agreement between the Colorado Coal and Iron Company and the Rail Makers' Association. It is logical to assume that the payment was for curtailing production of heavy rails in 1889.
[91] *Ibid.* See also *Pueblo Daily Chieftain*, May 4, June 5, 7, July 14, 1892.
[92] Colorado Fuel and Iron Company, "Predecessor Companies of The Colorado Coal & Iron Co.," p. 44.
[93] *Ibid.*
[94] *Ibid.*, p. 38.
[95] *Ibid.*, p. 44.
[96] Computed from Colorado Coal and Iron Company, *Annual Reports, 1880-1891.*
[97] Colorado Fuel and Iron Company, "Predecessor Companies of The Colorado Coal & Iron Co.," p. 30.
[98] Colorado Coal and Iron Company, *Annual Report, 1885*, pp. 13-14; *Annual Report, 1886*, p. 16; *Pueblo Daily Chieftain*, Nov. 5, 8, 22, Dec. 6, 1885.

[99] Colorado Fuel and Iron Company, "Predecessor Companies of The Colorado Coal & Iron Co.," p. 36; *Pueblo Daily Chieftain*, June 1, 1888, Jan. 20, 1889.

[100] Colorado Coal and Iron Company, *Annual Report, 1889* (1890), p. 16; Glenn W. Traer to John C. Osgood, Chicago, May 25, 1892.

[101] Colorado Coal and Iron Company Minute Book, III: 44, 75.

[102] *Ibid.*, III: 150-151; Colorado Fuel and Iron Company, "Predecessor Companies of The Colorado Coal & Iron Co.," p. 35.

[103] Colorado Fuel and Iron Company, "Predecessor Companies of The Colorado Coal & Iron Co.," p. 30.

[104] Colorado Coal and Iron Company, *Annual Report, 1890*, p. 12; *Annual Report, 1891*, p. 10.

[105] Colorado Fuel and Iron Company, "Predecessor Companies of The Colorado Coal & Iron Co.," pp. 21, 31.

[106] *Ibid.*

[107] *Ibid.*, p. 27; George S. Ramsay to E. M. Streck, Pueblo, Jan. 1, 1890.

[108] See mandate of U.S. Supreme Court, Nov. 21, 1887, in *U.S. vs. The Colorado Coal and Iron Company*, Docket 668; Colorado Coal and Iron Company, *Annual Report, 1887*, p. 16; *Engineering and Mining Journal*, XLVI (Nov. 26, 1887): 401.

[109] See lease, dated Feb. 27, 1888, Film O; Dougherty to Henry S. Grove, Pueblo, June 16, 1889, Film I; J. Imbrie Miller to DuPuy, Pueblo, Jan. 24, 1889; George S. Ramsay to E. M. Steck, Pueblo, Jan. 1, 1890.

[110] Dougherty to Grove, June 16, 1889. See also records for *U.S. vs. The Gray Creek Coal and Coking Company, and Michael Beshoar*, Docket 2311, U.S. Circuit Court, District of Colorado, in Federal Records Center, Region 8.

[111] The acquisition of this tract is related in Dougherty to Grove, Pueblo, June 18, 1889; Dougherty to Edward J. Berwind, Pueblo, June 20, 1889, June 21, 1889, Film I. See also Colorado Coal and Iron Company Minute Book, III: 153.

[112] Contract with W. A. Kearns, dated May 12, 1890, Film J; Colorado Fuel and Iron Company, "Monthly Tonnage and Cost Records, 1888-1912," in CF&I Steel Corporation, Mining Department, Pueblo.

[113] Colorado Fuel and Iron Company, "Predecessor Companies of The Colorado Coal & Iron Co.," p. 20.

[114] Colorado Coal and Iron Company, *Annual Report, 1883*, pp. 8-9. See also Colorado State Inspector of Coal Mines, *Annual Report, 1884*, p. 85.

[115] Colorado Fuel and Iron Company, "Predecessor Companies of The Colorado Coal & Iron Co.," p. 20.

[116] Ramsay to Steck, Jan. 1, 1891; Colorado Coal and Iron Company, *Annual Report, 1891*, p. 17; Colorado Fuel and Iron Company, "Monthly Tonnages and Cost, 1888-1912."

[117] Ramsay to Steck, Jan. 1, 1891.

[118] Colorado State Inspector of Coal Mines, *Biennial Report, 1885-1886* (Denver, 1887), p. 45; Colorado Fuel and Iron Company, "Monthly Tonnage and Costs, 1888-1912."

[119] Colorado Coal and Iron Company, *Annual Report, 1881*, p. 15; Colorado State Inspector of Coal Mines, *Biennial Report, 1885-1886*, p. 47; Colorado Fuel and Iron Company, "Predecessor Companies of The Colorado Coal & Iron Co.," p. 19.

[120] Colorado Coal and Iron Company, *Annual Report, 1881*, pp. 9-10; Colorado Fuel and Iron Company, "Predecessor Companies of The Colorado Coal & Iron Company," p. 20.

[121] Colorado State Inspector of Coal Mines, *Biennial Report, 1885-1886*, p. 57.

[122] Colorado Coal and Iron Company, *Annual Report, 1881*, p. 10.

[123] Colorado Fuel and Iron Company, "Predecessor Companies of The Colorado Coal & Iron Co.," p. 27.
[124] *Ibid.*
[125] Colorado Coal and Iron Company Minute Book, II: 308-309, 338-340. See also agreement, March 18, 1887, Film O.
[126] Colorado Coal and Iron Company Minute Book, III: 111. See agreement, dated April 18, 1889, Film I.
[127] *Pueblo Daily Chieftain*, July 18, Aug. 14, 1892.
[128] *Ibid.*, Jan. 28, 1891.

Chapter 7. The Colorado Fuel Company

[1] *Sketches of Colorado ... Being an Analytical Summary and Biographical History of the State of Colorado* (Denver, 1911), p. 227; *New York Times*, Sept. 7, 1902; *Rocky Mountain News*, Jan. 5, 1926.
[2] Brief A, p. 2, John L. Jerome Collection, The State Historical Society of Colorado, Denver. Cited hereafter as Jerome Brief.
[3] "Julian A. Kebler: Outline Sketch of a Life Full of Service," *Camp and Plant*, V (April 23, 1904): 353.
[4] "Estimates of Julian A. Kebler," *ibid.*, p. 360; *Denver Post*, Dec. 14, 1902.
[5] *Denver Post*, Dec. 14, 1902.
[6] *Sketches of Colorado*, p. 131; "The Late A. C. Cass," *Camp and Plant*, IV (Sept. 5, 1903): 186.
[7] Jerome Brief A, p. 1.
[8] *Sketches of Colorado*, pp. 272-273; *Portrait and Biographical Record of Denver and Vicinity, Colorado* (Chicago, 1898), pp. 1225-1226.
[9] Jerome Brief A, pp. 2-3.
[10] Colorado Fuel and Iron Company, "Predecessor Companies of The Colorado Coal & Iron Co.," p. 47.
[11] *Ibid.*, pp. 60-61.
[12] *Ibid.*, p. 48.
[13] For the origin of the Anthracite Mesa Mine, see Durango Trust, *Report to Subscribers, September 1, 1884* (Colorado Springs, 1884), pp. 13-14; "Hard Coal Mining at Anthracite," *Camp and Plant*, III (Jan. 28, 1903): 73. The Colorado Fuel Company's leases for the mine are on Film O.
[14] Colorado Fuel Company, *Annual Report, 1889* (New York, 1889), p. 5; Colorado State Inspector of Coal Mines, *Biennial Report, 1887-1888* (Denver, 1888), p. 101. See also Colorado Fuel and Iron Company, "Predecessor Companies of The Colorado Coal & Iron Co.," p. 49.
[15] Jerome Brief A, p. 3.
[16] Crystal River Toll Road Company Minute Book, pp. 2-35; Colorado Fuel and Iron Company, "Predecessor Companies of The Colorado Coal & Iron Co.," p. 53.
[17] Colorado and Utah Railway Company Minute Book, pp. 2-7; Colorado Fuel and Iron Company, "Predecessor Companies of The Colorado Coal & Iron Co.," p. 53.
[18] Denver Fuel Company Minute Book, pp. 1, 7.
[19] *Ibid.*, pp. 9, 22-23.
[20] Agreements between Sullivan and Sopris, April 27 and 28, 1887, Film G.
[21] *Trinidad Daily Advertiser*, Dec. 14, 1886. See also records in *U.S. vs. Charles D. Chase, Trustee, and Elbridge B. Sopris*, Docket 2176, U.S. Circuit Court, District of Colorado, in Federal Records Center, Region 8, Denver.
[22] Bill of complaint, Docket 2176.
[23] Copy of letter from Hobson to U.S. Attorney General, Denver, Feb. 25, 1888, Film G.
[24] *Ibid.*
[25] *Ibid.*

26 Colorado Fuel Company Minute Book, pp. 100-103; *Annual Report, 1889*, p. 6.

27 Colorado Fuel and Iron Company, "Predecessor Companies of The Colorado Coal & Iron Co.," p. 63.

28 Elk Mountain Fuel Company Minute Book, pp. 9-13.

29 *Ibid.*, pp. 23-24, 37-38; Agreement between J. C. Osgood and Elk Mountain Fuel Company, dated Nov. 7, 1888, Film G.

30 Colorado Fuel and Iron Company, "Predecessor Companies of The Colorado Coal & Iron Co.," p. 48.

31 Colorado Fuel Company, *Annual Report, 1889*, p. 1.

32 Colorado Fuel Company Minute Book, pp. 37, 50-51. The bill of sale, dated April 1, is on Film G.

33 Colorado Fuel Company Minute Book, p. 87. See also bill of sale, dated April 25, 1889, and Elk Mountain Fuel Company Minutes of Stockholders' Meeting, Aug. 29, 1889, Film G.

34 Colorado Fuel and Iron Company, "Predecessor Companies of The Colorado Coal & Iron Co.," pp. 51-52.

35 *Ibid.*, pp. 49-50.

36 Southern Land Company Minute Book, pp. 1-5. See also Colorado Fuel and Iron Company, "Predecessor Companies of The Colorado Coal & Iron Co.," p. 69.

37 Southern Land Company Minute Book, pp. 7-8; lease to the Colorado Coal and Iron Company, dated Dec. 20, 1888, Film O; assignment of leases to the Colorado Fuel Company, dated Nov. 28, 1889, Film G.

38 Agreement with Danforth, Brown and Cameron, dated Sept. 28, 1889, Film J; Agreement with Whitney, Joy, and Dunbaugh, dated Oct. 3, 1889, Film G. See also Southern Land Company Minute Book, pp. 31-36.

39 Colorado Fuel Company Minute Book, p. 135. See also Coppell to Beaman, New York, Sept. 10, 20, 1889, Film G.

40 Colorado Fuel and Iron Company, "Predecessor Companies of The Colorado Coal & Iron Co.," p. 68; Colorado State Inspector of Coal Mines, *Biennial Report, 1887-1888*, pp. 83, 87; *Pueblo Daily Chieftain*, Aug. 3, 9, Dec. 5, 1888.

41 Colorado Fuel Company Minute Book, pp. 95, 125, 134-137.

42 *Ibid.*, pp. 155-158.

43 Receipts for final settlement, dated April 20, 1893, Film F. See also Colorado Fuel and Iron Company, "Predecessor Companies of The Colorado Coal & Iron Co.," p. 70.

44 Colorado Fuel and Iron Company, "Predecessor Companies of The Colorado Coal & Iron Co.," p. 52.

45 Colorado Fuel Company Minute Book, pp. 85, 169.

46 "Anthracite Mine at Floresta." *Camp and Plant*, IV (Aug. 8, 1903): 77-78.

47 Old Recorded Deeds, File 52, Nos. 24 and 25, Land and Water Department, Pueblo.

48 Denver Fuel Company Minute Book, pp. 73-74, 95-97, 101-102. See also subscription list, dated Feb. 15, 1890, Film G.

49 Old Recorded Deeds, File 54, No. 1.

50 Denver Fuel Company Minute Book, p. 74.

51 Old Recorded Deeds, File 54, No. 2; agreement between Osgood and the Denver Fuel Company, dated April 1, 1890, Film G; Colorado Fuel Company, *Annual Report, 1890* (New York, 1890), p. 8.

52 Yampa River Coal Company Minute Book, pp. 2-7.

53 Agreement, dated Feb. 13, 1890, Film G.

54 Denver Fuel Company Minute Book, pp. 83-84. Assignments by Choate, Wood and Osgood appear on Film K. See also Colorado Fuel and Iron Company, "Predecessor Companies of The Colorado Coal & Iron Co.," p. 54.

[55] Colorado State Inspector of Coal Mines, *Annual Report, 1884*, pp. 61-62.

[56] R. M. Hosea to A. C. Cass, Denver, June 8, 1891, Letterpress, Engineering Section, Pueblo. See also Colorado State Inspector of Coal Mines, *Biennial Report, 1887-1888*, p. 101; *Biennial Report, 1889-1890*, p. 62; Colorado Fuel and Iron Company, "Predecessor Companies of The Colorado Coal & Iron Co.," p. 49.

[57] Colorado Fuel Company Minute Book, pp. 101-103; *Annual Report, 1889*, p. 6. See also *Engineering and Mining Journal*, XLVI (Oct. 6, 1888): 289.

[58] Colorado State Inspector of Coal Mines, *Biennial Report, 1887-1888*, p. 83; "The Colorado Supply Company." *Camp and Plant*, V (March 26, 1904): 241-242.

[59] Colorado State Inspector of Coal Mines, *Biennial Report, 1887-1888*, pp. 55-56; *Biennial Report, 1889-1890*, pp. 41-42.

[60] Colorado Fuel and Iron Company, "Monthly Tonnages and Costs, 1888-1912."

[61] Miller, "History of Coal Development in Colorado, 1880-1900," p. 26.

[62] Las Animas and Western Railway Company Minute Book, pp. 3-10. See also Colorado Fuel and Iron Company, "Predecessor Companies of The Colorado Coal & Iron Co.," pp. 53-54.

[63] Colorado Fuel Company, *Annual Report, 1890*, p. 4.

[64] Colorado Fuel and Iron Company, "Predecessor Companies of The Colorado Coal & Iron Co.," p. 49.

[65] "Pictou, Colorado: A Rich Mine of Domestic Coal." *Camp and Plant*, I (June 21, 1902): 529.

[66] Colorado Fuel and Iron Company, "Predecessor Companies of The Colorado Coal & Iron Company," p. 72. See also correspondence and agreement between Kebler and Frank Trumbull concerning sale of coal to the Union Pacific, Film J.

[67] Agreement between Wheeler and Hagerman, Feb. 25, 1886; John J. Lipsey, *The Lives of James John Hagerman, Builder of The Colorado Midland Railway* (Denver, 1968), pp. 31, 36-38.

[68] Certificate of Incorporation, Film I.

[69] Grand River Coal and Coke Company Minute Book of Trustees, pp. 6-7.

[70] *Ibid.*, p. 15.

[71] *Ibid.*, pp. 11-13. See also Lipsey, *The Lives of James John Hagerman*, pp. 66, 71.

[72] Agreement between the Colorado Coal and Iron Company and the Grand River Coal and Coke Company, March 18, 1887, Film O; Grand River Coal and Coke Company Minute Book of Board of Trustees, pp. 23-24; W. B. Devereaux to President and Board of Trustees, Glenwood Springs, Dec. 31, 1887, Film I. See also "Sunlight and Gulch." *Camp and Plant*, I (April 12, 1902): 289-290.

[73] Colorado Fuel and Iron Company, "Predecessor Companies of The Colorado Coal & Iron Co.," p. 72; *Engineering and Mining Journal*, XLV (Jan. 21, 1888): 58; XLVI (Sept. 28, 1888): 267.

[74] Colorado Fuel and Iron Company, "Predecessor Companies of The Colorado Coal & Iron Co.," pp. 72-73; Colorado State Inspector of Coal Mines, *Biennial Report, 1887-1888*, p. 95; *Biennial Report, 1889-1890*, pp. 57-58; Grand River Coal and Coke Company, *Annual Report, 1891* (1892), p. 8.

[75] Grand River Coal and Coke Company, "Statement of Account, December 31, 1891," Film I.

[76] Agreements with the Kansas and Colorado Company, March 12, Aug. 6, 1889, in Grand River Coal and Coke Company Minute Book of Trustees, pp. 52-54, 71-73.

[77] Miller, "History of Coal Development in Colorado, 1880-1900," p. 26; *Engineering and Mining Journal*, XLVI (July 30, 1887): 83; Grand River Coal and Coke Company, *Annual Report, 1891*, p. 10; J. A. Kebler to J. C. Osgood, Denver, May 21, 1892.

[78] Kebler to Osgood, May 21, 1892.

[79] *Ibid.* See also Devereaux to Wheeler, Glenwood Springs, Aug. 2, 1888, Aug. 13, 1889, Film I.

[80] Grand River Coal and Coke Company, "Statement of Account, Dec. 31, 1891."

[81] Miller, "History of Coal Development in Colorado, 1880-1900."

[82] Kebler to Osgood, May 21, 1889.

[83] Agreement, dated June 8, 1892, Film G.

[84] Grand River Coal and Coke Company Minute Book of Trustees, pp. 113-117.

[85] Colorado Fuel and Iron Company, "Predecessor Companies of The Colorado Coal & Iron Co.," p. 50.

[86] *Ibid.*, pp. 56-57.

Chapter 8. The Resurgence of the Steel Works, 1892-1903

[1] H. R. Wolcott and J. C. Osgood to Stockholders of the Colorado Fuel Company, Denver, Sept. 2, 1892, Film G.

[2] *Rocky Mountain News*, March 25, 1892.

[3] F. A. Delano to Paul Morton, Chicago, July 12, 1892. See also Wolcott and Osgood to Stockholders of the Colorado Fuel Company, Sept. 2, 1892; Delano to Osgood, Chicago, May 27, 1892; E. A. Kebler to J. A. Kebler, Denver, June 17, 1892.

[4] *Pueblo Daily Chieftain*, June 13, 1892; *Rocky Mountain News*, June 17, 1892.

[5] Jerome Brief A, p. 6.

[6] *Ibid.*

[7] "Memorandum of reasons why it would be to the advantage of The Colorado Coal & Iron Company, as well as the consolidated company that its management should be conducted by Mr. Osgood, pending the final and legal consolidation of the companies."

[8] *Pueblo Daily Chieftain*, Aug. 1, 2, 3, 1892.

[9] *Ibid.*, Aug. 4, 5, 1892.

[10] Jerome Brief A, p. 7.

[11] The plan of consolidation is in Colorado Coal and Iron Company Minute Book, III: 277-284.

[12] Berwind, Meek, Thalmann, and McHarg to Stockholders of the Colorado Coal and Iron Company, New York, Sept. 1, 1892, Film H.

[13] *Ibid.*

[14] *Ibid.*

[15] Wolcott and Osgood to Stockholders of the Colorado Fuel Company, Sept. 2, 1892.

[16] *Ibid.* On Sept. 9, 1891, the Colorado Fuel Company divided its capital stock into $2,000,000 preferred and $3,000,000 common shares. See Colorado Fuel Company Minute Book, pp. 210-212.

[17] Colorado Coal and Iron Company Minute Book, III: 279-280.

[18] *Ibid.*, 280.

[19] Colorado Fuel and Iron Company Minute Book of Directors, pp. 5-16. That same day, there were three resignations from the board: Arthur was succeeded by E. J. Berwind, C. K. McHarg was succeeded by Ernest Thalmann, and Curtis was succeeded by H. K. McHarg.

[20] *Ibid.*, pp. 27, 32. Toll was elected second vice-president but resigned the same day. He was succeeded by Morton.

[21] Colorado Fuel and Iron Company, *Annual Report, 1893* (Denver, 1893), p. 4.

[22] Wolcott and Osgood to Stockholders of the Colorado Fuel Company, Sept. 2, 1892.

[23] *Ibid.* See also Colorado Fuel and Iron Company, "Predecessor Companies of The Colorado Coal & Iron Co.," p. 81; *Annual Report, 1892,* p. 17.

[24] Colorado Fuel and Iron Company, *Annual Report, 1893,* pp. 16-17. The capacities of these units were reported at considerably lower figures a year earlier by Wolcott and Osgood to Stockholders of the Colorado Fuel Company, Sept. 1892.

[25] Wolcott and Osgood to Stockholders of the Colorado Fuel Company, Sept. 2, 1892.

[26] *Pueblo Daily Chieftain,* Nov. 1, 1892, Jan. 1, 1893; *New York Iron Age,* June 9, 1932.

[27] *Pueblo Daily Chieftain,* Dec. 7, 1899, May 17, 1906; *Camp and Plant,* IV (Aug. 15, 1903): 110.

[28] Agreement between the Colorado Fuel and Iron Company and Crowe, Oct. 25, 1892, Film K; *Pueblo Daily Chieftain,* Nov. 20, 1892, Jan. 1, 1893; *The Blast,* March 3, 1932.

[29] Agreement between the Colorado Fuel and Iron Company and Brown, Nov. 7, 1892, Film K; *Pueblo Daily Chieftain,* Nov. 20, 1892, Jan. 1, 1893; *Camp and Plant,* IV (Aug. 15, 1903): 113-114; *The Blast,* Sept. 29, 1930, Oct. 23, 1936.

[30] *Camp and Plant,* IV (Aug. 15, 1903): 110.

[31] Colorado Fuel and Iron Company, *Annual Report, 1893,* p. 7; Jerome Brief A, p. 11.

[32] Jerome Brief A, pp. 11-12.

[33] *Ibid.*, Brief B, p. 1.

[34] Colorado Fuel and Iron Company, *Annual Report, 1894* (Denver, 1894), p. 5.

[35] *Ibid.*, p. 13; *Pueblo Daily Chieftain,* Aug. 31, 1893.

[36] Colorado Fuel and Iron Company, *Annual Report, 1894,* p. 7; *Engineering and Mining Journal,* LVIII (Sept. 1, 1894): 204.

[37] Colorado Fuel and Iron Company, *Annual Report, 1894,* pp. 10-11.

[38] Jerome Brief B, p. 2. See also Colorado State Inspector of Coal Mines, *Biennial Report, 1893-1894* (Denver, 1895), pp. 4-5.

[39] Colorado Fuel and Iron Company, *Annual Report, 1895* (Denver, 1895), p. 3; "Predecessor Companies of The Colorado Coal & Iron Co.," p. 117.

[40] Jerome Brief B, pp. 4-5; Osgood to Jerome, New York, July 6, 1895, in Jerome Collection.

[41] Jerome Brief B, pp. 14-16; Osgood to Jerome, New York, July 19, 27, 1895, in Jerome Collection: Colorado Fuel and Iron Company Minute Book of Directors, pp. 60-61.

[42] Jerome Brief B, p. 2; Colorado Fuel and Iron Company, "Predecessor Companies of The Colorado Coal & Iron Co.," p. 148.

[43] *Engineering and Mining Journal,* LXIII (Feb. 13, 1897): 173.

[44] Agreement between Robert F. Kennedy and Osgood, Film O.

[45] Agreement between Luther S. Rent and Osgood, Dec. 16, 1896, Film O.

[46] Agreement between J. W. Gates and Osgood, Dec. 19, 1896, Film L.

[47] *Engineering and Mining Journal,* LXIII (Feb. 13, 1897): 173; (Feb. 20, 1897): 182.

[48] Jerome Brief B, p. 18; Brief C, p. 3; Colorado Fuel and Iron Company, *Annual Report, 1897* (Denver, 1897), pp. 5-9; *Annual Report, 1898* (Denver, 1898), p. 4.

[49] "Minutes of the Meeting of Rail Makers," Sept. 22, 26, 27, Oct. 4, 1898; Osgood to Cass, Sept. 23, 28, 1898.

[50] "Memorandum of Agreement, November 2, 1898," signed by Powell Stackhouse, J. C. Osgood, C. C. Cluff, and C. W. Baker as the New York Committee.

51 Osgood to Cass, New York, Nov. 20, 1898.

52 *Ibid.*

53 *Ibid.*

54 *Ibid.*

55 "Memorandum of November 28, 1898." See also Osgood to Cass, Nov. 26, 1898; W. H. Vilas to members of the New York Committee, Dec. 16, 1898.

56 "Memorandum of percentage allowed each interest in Rail Association, which commenced business in October, 1898."

57 *Ibid.* "Memorandum of understanding between J. C. Osgood, of one part, and E. H. Gary and others, of other part, October 1, 1900." See also Osgood to Cass, New York, Oct. 10, 1900.

58 Colorado Fuel and Iron Company, "Predecessor Companies of The Colorado Coal & Iron Company," p. 48.

59 *Ibid.*

60 Colorado Fuel and Iron Company Minute Book of Directors, p. 77; *Annual Report, 1897,* p. 8; *Engineering and Mining Journal,* LXIII (Jan. 16, 1897): 71.

61 Colorado Fuel and Iron Company, *Annual Report, 1898,* pp. 4-5; Osgood to Stockholders of The Colorado Fuel and Iron Company, New York, Sept. 5, 1899.

62 Osgood to Stockholders, Sept. 5, 1899.

63 *Ibid.* See also Colorado Fuel and Iron Company, *Annual Report, 1900* (Denver, 1900), pp. 4-5.

64 Jerome Brief C, pp. 5-8, Brief D, p. 14.

65 *Ibid.*, Brief C, pp. 5-6.

66 *Ibid.*, Brief G, p. 5. See also bill of complaint, John L. Jerome vs. John C. Osgood and the Colorado Finance and Construction Company, Jerome Collection. This document was prepared for a suit that was not instituted because of Jerome's death in 1903. Cited hereafter as "bill of complaint, Jerome vs. Osgood."

67 See articles of incorporation in Jerome Brief G-2, pp. 2-4.

68 See bill of complaint, Jerome vs. Osgood.

69 Colorado Finance and Construction Company Minutes, Oct. 4, 1899, in Jerome Brief G-2, p. 7; Colorado Fuel and Iron Company Minute Book of Directors, pp. 120-121; *Annual Report, 1901* (Denver, 1901), p. 23.

70 Jerome Brief G, p. 7.

71 Colorado Fuel and Iron Company, *Annual Report, 1901,* p. 23.

72 Jerome Brief G, p. 8.

73 *Ibid.*, Brief I, p. 4; Brief J, p. 16.

74 *Ibid.*, Brief I, p. 5.

75 *Ibid.*, Brief J, p. 14.

76 *Ibid.*, Brief I, p. 6.

77 Osgood to Jerome, New York, March 19, 1900, Jerome Collection.

78 Colorado Finance and Construction Company Minutes of Directors, April 19, 1900, in Jerome Brief G-2, p. 10. See also Brief J, p. 14.

79 Jerome Brief J, pp. 16-17.

80 *Ibid.*, pp. 17, 21.

81 Jerome to Osgood, Aug. 28, 1899, in Jerome Collection.

82 Colorado Fuel and Iron Company, *Annual Report, 1896* (Denver, 1896), p. 7; *Annual Report, 1898,* p. 6. See also pertinent records in *Henry K. Pomeroy and Cornelius B. Gold vs. The Colorado Coal and Iron Development Company,* Docket 3492, U.S. Circuit Court, District of Colorado, in Federal Records Center, Region 8, Denver.

83 Pueblo Realty Trust Company Minute Book, pp. 2-7.

[84] *Commercial and Financial Chronicle*, LIX (Nov. 3, 1894): 780, (Dec. 29, 1894): 1141.
[85] Jerome Brief I, pp. 8, 15.
[86] Bill of complaint, Jerome vs. Osgood.
[87] Jerome Brief I, pp. 15-16.
[88] Minnequa Town Company, *Annual Report, 1901* (Pueblo, 1901), p. 3. See also *Camp and Plant*, IV (Aug. 15, 1903): 110-114.
[89] Jerome to Osgood, April 4, 1900, in Brief I, pp. 17-19; Minnequa Town Company, *Annual Report, 1901*, p. 3; *Pueblo Daily Chieftain*, May 24, Aug. 11, Oct. 11, 1900.
[90] Colorado Fuel and Iron Company Minute Book of Directors, pp. 137-138; Osgood to Stockholders, June 6, 1901, in Colorado Fuel and Iron Company *Annual Report, 1901*, pp. 23-24. See also Colorado Finance and Construction Company Minutes, April 4, 1901, in Jerome Brief G-2, pp. 20-22.
[91] For information about the Rocky Mountain Coal and Iron Company and its acquisition of the Maxwell lands, see the corporation's Minute Book, pp. 21-25.
[92] Colorado Fuel and Iron Company Minutes of Directors, p. 139. Previous attempts to curb competition between the two firms is revealed in A. C. Cass to Osgood, Jan. 31, Feb. 14, 1899, Letterpress, Engineering Section; Agreement between the Colorado Fuel and Iron Company and the Laramie Iron and Steel Company, Feb. 20, 1899, Film O.
[93] Bill of complaint, Jerome vs. Osgood.
[94] *Pueblo Daily Chieftain*, May 24, 1900.
[95] *Ibid.*
[96] Osgood to Stockholders, June 6, 1901, in Colorado Fuel and Iron Company, *Annual Report, 1901*, p. 25. See also *Pueblo Daily Chieftain*, May 18, 1901.
[97] Osgood to Jerome, New York, April 20, 1901, Jerome Collection.
[98] For Gates' business career, see Lloyd Wendt and Herman Kogan, *Bet a Million! The Story of John W. Gates* (Indianapolis and New York, 1948), *passim.*
[99] Osgood to Jerome, New York, May 10, 1901, Jerome Brief O, p. 33.
[100] *Ibid.*
[101] Colorado Fuel and Iron Company Minute Book of Stockholders, pp. 23-73; Minute Book of Directors, pp. 217, 244-245; *Annual Report, 1901*, p. 5.
[102] Colorado Fuel and Iron Company Minute Book of Directors, pp. 244-245.
[103] *Ibid.*, pp. 245-248.
[104] *Ibid.*, p. 253.
[105] Colorado Fuel and Iron Company, "Predecessor Companies of The Colorado Coal & Iron Co.," p. 96.
[106] *Pueblo Daily Chieftain*, Aug. 24, 1901.
[107] For a comparison of the steel works in 1892 and 1903, see "Steel Works in Old Times: Resume of the History of Minnequa Works." *Camp and Plant*, II (Oct. 4, 1902): 321-326, and "The Minnequa Works of The Colorado Fuel and Iron Company." V (Feb. 27, 1904): 149-155.
[108] "Steel Works in Old Times: Resume of the History of Minnequa Works," p. 325.
[109] Colorado Fuel and Iron Company, "Predecessor Companies of The Colorado Coal & Iron Co.," p. 101.
[110] Descriptions of these facilities are in J. A. Durfee, "Open Hearth Process for Steel Making." *Camp and Plant*, IV (Oct. 31, 1903): 366-367; and T. J. Brown, "Blooms, Billets and Bars: History of Evolution of Rolling Merchant Steel and Description of New Mills of Merchant Steel Department at Minnequa Works." IV (Aug. 22, 1903): 125-135.
[111] See "The Minnequa Works of The Colorado Fuel and Iron Company," pp. 152-154; John Y. Brooks, "Wire Making and the C. F. & I. Wire Mill." *Camp*

and Plant, III (Jan. 17, 1903): 49-62; A. L. Conner, "The Tin Plate and Sheet Steel Department." III (May 9, 1903): 411-416; "Rolling Rods." IV (Oct. 3, 1903): 269-271.

[112] See R. M. Hosea, "Minnequa Steel Works Water Supply." *Camp and Plant*, II (Nov. 29, 1902): 513-519; W. E. Berg, "Water System of the Minnequa Works." *Colorado Fuel & Iron Company Industrial Bulletin*, X (Oct. 15, 1925): 19-20.

[113] J. B. Stone, "Calumet Iron Tract, Chaffee County, August 19, 1933," in Mining Department, Pueblo.

[114] John B. Stone, "History of the CF&I Ore Supply." *The Blast*, March 4, 1932. See also "Orient Production," in Mining Department, Pueblo.

[115] Stone, "History of the CF&I Ore Supply." See also R. C. Hills, "Reference Synopsis of the Known Iron Ore Deposits of the Rocky Mountains and Pacific Coast," in Hills "Reports on Iron Properties," pp. 1-78.

[116] "Iron Mining in Colorado," pp. 5, 9-10, 14, in Mining Department, Pueblo.

[117] See Hills "Reports on Iron Properties," *passim.*

[118] *Ibid.*, pp. 202-226.

[119] Colorado Fuel and Iron Company Minute Book of Directors, pp. 108-109; "Predecessor Companies of The Colorado Coal & Iron Co.," p. 109.

[120] "Fierro and the Fierro Mines." *Camp and Plant*, I (May 24, 1902): 434-435; *Engineering and Mining Journal*, LXVIII (Dec. 30, 1899): 798.

[121] *Engineering and Mining Journal*, LXX (Oct. 27, 1900): 499; *Pueblo Daily Chieftain*, Sept. 17, 1900.

[122] E. P. Snow, "The Hartville Iron Ore Deposits in Wyoming," *Engineering and Mining Journal*, LX (Oct. 5, 1895): 320-321.

[123] Ramsay to E. M. Steck, Pueblo, May 30, 1890.

[124] *Engineering and Mining Journal*, LV (April 22, 1893): 376.

[125] Hills, "Reports on Iron Properties," pp. 247-259.

[126] Contract, Film P.

[127] *Engineering and Mining Journal*, LXX (Oct. 27, 1900): 500. See also "Production of Sunrise Mines" and "History of The Sunrise Mines," in Mining Department, Pueblo. Open-pit mining was abandoned in 1942, in part because grades had become too steep for locomotive haulage.

[128] *Pueblo Daily Chieftain*, Feb. 5, March 5, 1900.

[129] "The Colorado and Wyoming Railway," p. 145. See also Thomas Tucker, "Up-To-Date Information About the C. & W., Northern Division." *Colorado Fuel & Iron Company Industrial Bulletin*, X (June 15, 1925): 6-7.

[130] Colorado Fuel and Iron Company Minute Book of Directors, p. 229.

[131] Agreement with Hartville Iron Company, Jan. 27, 1910.

[132] "List of Iron Lands, June 22, 1901," in Mining Department, Pueblo. See also F. J. Hearne to F.T. Gates, April 14, 1904; Hearne to G. W. McClure, Son & Co., May 2, 1904.

[133] "Construction and Improvements at C. F. & I. Co. Plants." *Camp and Plant*, II (Nov. 15, 1902): 470; "The Little Grand Manganese Mine." III (May 16, 1903): 437-439; "Edwin Thatcher Wolverton." IV (Aug. 15, 1903): 111.

[134] "The Little Grand Manganese Mine," p. 438; *Engineering and Mining Journal*, LXXIII (Feb. 1, 1903): 192.

[135] Hearne to F. T. Gates, March 17, 1904.

[136] Hills, "Reports on Iron Properties," pp. 90-94.

[137] See lease from John Hed to C. B. Saxton and L. L. Harding, Sept. 4, 1895, assigned to the Colorado Fuel and Iron Company Nov. 10, 1897, Film L; "The Limestone Quarry at Lime," p. 2.

[138] "The Limestone Quarry at Lime," p. 3.

[139] Colorado Fuel and Iron Company Minute Book of Directors, p. 310.

[140] R. B. Jerrard, "Calcite Quarries." *Colorado Fuel & Iron Company Industrial Bulletin*, X (April 5, 1925): 7-8.

Chapter 9. Expansion of the Fuel Trade, 1892-1903.

[1] Colorado Fuel and Iron Company, "Predecessor Companies of The Colorado Coal & Iron Co.," p. 149.

[2] R. C. Hills, "Reports on Coal Properties," p. 175.

[3] R. C. Hills, *Coal and Iron Lands of The Colorado Fuel and Iron Company (Condensed Report)* (Denver, 1893), p. 10; T. A. Rickard, "Across the San Juan Mountains." *Engineering and Mining Journal*, LXXVI (Sept. 19, 1903): 423.

[4] Rickard, "Across the San Juan Mountains," pp. 423-424; "Anthracite Mine at Floresta." *Camp and Plant*, IV (Aug. 8, 1903): 77; Hills, *Coal and Iron Lands of the Colorado Fuel and Iron Company (Condensed Report)*, p. 10; Miller, "History of Coal Development in Colorado, 1880-1900," p. 26.

[5] Rickard, "Across the San Juan Mountains," pp. 423-424.

[6] Crystal River Railway Company Minute Book, pp. 5-11.

[7] *Ibid.*, p. 25; Crystal River Toll Road Company Minute Book, p. 33; Colorado and Utah Railway Company Minute Book, pp. 45-49; Aspen and Western Railway Company Minute Book, pp. 61-62. See also Colorado Fuel and Iron Company, "Predecessor Companies of The Colorado Coal & Iron Co.," pp. 34-35.

[8] Crystal River Railway Company Minute Book, p. 31.

[9] Hills, "Reports on Coal Properties," pp. 611-612, 614, 623.

[10] Jim Berry Pearson, *The Maxwell Land Grant* (Norman, 1961), p. 229.

[11] Hills, "Reports on Coal Properties," pp. 311-314.

[12] Pearson, *The Maxwell Land Grant*, p. 229.

[13] Colorado Fuel and Iron Company, *Annual Report, 1894*, p. 10.

[14] *Ibid.*, pp. 10-11; *Pueblo Daily Chieftain*, May 1, June 20, 1894.

[15] Colorado Fuel and Iron Company, *Annual Report, 1894*, p. 11.

[16] *Pueblo Daily Chieftain*, June 14, 1894.

[17] Colorado State Inspector of Coal Mines, *Biennial Report, 1893-1894*, pp. 5-6.

[18] *Pueblo Daily Chieftain*, May 25, 30, June 5, 6, 7, 9, 1894.

[19] *Ibid.*, June 20, 21, 22, 23, 24, 1894.

[20] *Ibid.*, June 24, 1894.

[21] *Engineering and Mining Journal*, LVIII (Aug. 4, 1894): 109; Oct. 6, 1894): 326.

[22] "The Colorado Supply Company," pp. 241-242.

[23] *Ibid.*, p. 245.

[24] *Pueblo Daily Chieftain*, June 14, 24, 1894.

[25] Lease, Dec. 5, 1895, Film L.

[26] Colorado Fuel and Iron Company, "Monthly Tonnages and Costs, 1888-1912."

[27] Colorado Fuel and Iron Company, *Annual Report, 1896* (Denver, 1896), p. 8; Minute Book of Directors, pp. 74-75.

[28] Osgood to C. M. Higginson, Denver, Sept. 3, 1896. See also E. P. Ripley to A. F. Walker, Chicago, Sept. 22, 1899, in AT&SF Collection, Kansas State Historical Society, Topeka.

[29] Contract with Cerrillos Coal Railroad Company, Film R; J. A. Writer to Osgood, Denver, Feb. 18, 1903, Film E.

[30] Contract with Oak Creek Coal Company, June 10, 1892, Film I; Colorado Fuel and Iron Company, "Predecessor Companies of The Colorado Coal & Iron Co.," pp. 182-183.

[31] "Rouse and Hezron." *Camp and Plant*, I (March 8, 1902): 193; Colorado Fuel and Iron Company, "Predecessor Companies of The Colorado Coal & Iron Co.," pp. 177-179.

[32] Colorado State Inspector of Coal Mines, *Biennial Report, 1897-1898* (Denver, 1899), pp. 77-78.

33 Ibid., *1899-1900* (Denver, 1901), p. 73; Colorado Fuel and Iron Company, "Predecessor Companies of The Colorado Coal & Iron Co.," p. 187.

34 *Camp and Plant*, I (Jan. 1, 1902): 7-11; Colorado Fuel and Iron Company, "Predecessor Companies of The Colorado Coal & Iron Co.," pp. 190-191.

35 "Coalbasin, Colorado." *Camp and Plant*, II (Sept. 13, 1902): 250; Colorado State Inspector of Coal Mines, *Biennial Report, 1899-1900*, p. 84; *Pueblo Daily Chieftain*, Dec. 12, 1900.

36 Colorado Fuel and Iron Company Comptroller's Office, "The Crystal River Railroad Company," March 9, 1929, typescript.

37 "The Crystal River Railroad." *Camp and Plant*, III (Jan. 10, 1903): 26; *Pueblo Daily Chieftain*, Dec. 12, 1900; *Denver Republican*, July 25, 1903.

38 Colorado Fuel and Iron Company, *Annual Report, 1902* (Denver, 1902), p. 5; "Predecessor Companies of The Colorado Coal & Iron Co.," pp. 150, 157-158, 160-161, 163-166.

39 "Berwind and Tabasco." *Camp and Plant*, I (Jan. 11, 1902): 57-59; Colorado Fuel and Iron Company, "Monthly Tonnages and Costs, 1888-1912."

40 "Rouse and Hezron," p. 196; Mining Department File 141-763.

41 Pearson, *The Maxwell Land Grant*, p. 232; Colorado Fuel and Iron Company Minute Book of Directors, pp. 137-138.

42 "Primero, Colorado." *Camp and Plant*, II (Oct. 25, 1902): 326; Colorado Fuel and Iron Company, *Annual Report, 1902*, p. 5; Colorado State Inspector of Coal Mines, *Biennial Report, 1901-1902* (Denver, 1903), p. 93.

43 "Tercio, Colorado." *Camp and Plant*, II (Aug. 13, 1902): 130; Colorado Fuel and Iron Company, "Predecessor Companies of the Colorado Coal & Iron Co.," p. 164.

44 "Cuatro, Colorado." *Camp and Plant*, IV (Nov. 28, 1903): 464; Colorado Fuel and Iron Company, "Predecessor Companies of The Colorado Coal & Iron Co.," pp. 165-166; Mining Department File 141-763.

45 Colorado Fuel and Iron Company Minute Book of Directors, pp. 216-217; Colorado and Wyoming Railway Company, *Mortgage to the Mercantile Trust Company, March 1, 1903, Securing First Mortgage Four Per Cent. Fifty-Year Bonds* (1903), pp. 1-2; "The Colorado and Wyoming Railway." *Camp and Plant*, III (Feb. 18, 1903): 145-146; C. J. Lee, "Historic Facts Concerning the C. & W., Southern Division," *Colorado Fuel & Iron Company Industrial Bulletin*, X (June 15, 1925): 16.

46 Colorado and Wyoming Railway Company, *Mortgage to The Mercantile Trust Company, March 1, 1903*, pp. 1-2.

47 "The Colorado and Wyoming Railway," p. 146.

48 *Ibid.* See also *Camp and Plant*, III (Feb. 18, 1903): 160.

49 "The Colorado and Wyoming Railway," pp. 148-149.

50 See Colorado State Inspector of Coal Mines, *Biennial Report, 1893-1894*, insert front cover; *Biennial Report, 1903-1904* (Denver, 1905), *passim.*

51 *Ibid.*

52 Colorado Fuel and Iron Company Minute Book of Directors, pp. 100-101; "Predecessor Companies of The Colorado Coal & Iron Co.," p. 197; Ripley to Victor Morawetz, Chicago, June 9, 1905, AT&SF Collection.

53 *Camp and Plant*, I (Dec. 28, 1901): 31; "The Gallup District of New Mexico." I (March 29, 1902): 242; *Engineering and Mining Journal*, LXXIII (March 1, 1902): 313.

54 Colorado State Inspector of Coal Mines, *Biennial Report, 1903-1904*, p. 47.

55 *Ibid.*

56 Colorado Fuel and Iron Company, "Predecessor Companies of The Colorado Coal & Iron Co.," pp. 152-155.

57 "Segundo, Colorado, and the Segundo Coke Ovens." *Camp and Plant*, III (April 15, 1903): 364, 366.

[58] L. B. Merriam, "The Development of a New Coal Field in Colorado." *Camp and Plant,* V (Feb. 20, 1904): 125-131; Colorado Fuel and Iron Company, "Predecessor Companies of The Colorado Coal & Iron Co.," pp. 164-165.
[59] "Berwind and Tabasco," pp. 57-58; Colorado Fuel and Iron Company, "Predecessor Companies of The Colorado Coal & Iron Co.," p. 158.
[60] "Cardiff and the Cardiff Ovens." *Camp and Plant,* I (May 10, 1902): 386.
[61] Colorado Fuel and Iron Company, "Predecessor Companies of The Colorado Coal & Iron Co.," p. 191; "Monthly Tonnages and Costs, 1888-1912."
[62] Colorado Bureau of Labor Statistics, *Biennial Report, 1899-1900* (Denver, 1900), p. 186.
[63] *Ibid.,* p. 187.
[64] *Ibid.,* p. 169; J. A. Kebler to J. E. Lundstrom, Denver, June 2, 1899, in William A. Bell Collection.
[65] Colorado Bureau of Labor Statistics, *Biennial Report, 1899-1900,* p. 196.
[66] Report of the joint committee to the General Assembly, Feb. 25, 1901, in Colorado Bureau of Labor Statistics, *Biennial Report, 1901-1902* (Denver, 1902), pp. 149-154.
[67] *Ibid.,* p. 155.
[68] *Ibid.; Pueblo Daily Chieftain,* March 24, April 2, 1901.
[69] Colorado Bureau of Labor Statistics, *Biennial Report, 1901-1902,* p. 153.
[70] *Ibid.,* 1903-1904 (Denver, 1904), p. 184; *Engineering and Mining Journal,* LXXVI (Nov. 7, 1903); 710.
[71] *Rocky Mountain News,* March 24, June 12, Oct. 13, 1904.

Chapter 10. Pioneering in Industrial Medicine

[1] For biographical data, see "Personnel of the Medical Department." *Camp and Plant,* II (Aug. 6, 1902): 165; Smiley, ed., *Semi-Centennial History of the State of Colorado,* II: 235-237; Colorado Fuel and Iron Company, *Annual Report of the Medical Department, 1929* (Pueblo, 1929), pp. 3-11.
[2] "Memories of Early Days in Minnequa," p. 13.
[3] "The Medical Department of The Colorado Fuel and Iron Company." *Camp and Plant,* I (Feb. 1, 1902): 105-105; "Chief Surgeon's 47th Annual Report." *Colorado Fuel & Iron Company Industrial Bulletin,* XIII (Sept. 1928): 4-5.
[4] "Chief Surgeon's 47th Annual Report," pp. 4-5; *Pueblo Daily Chieftain,* April 23, 1882.
[5] "The Medical Department of The Colorado Fuel and Iron Company," p. 106.
[6] Danforth to D. C. Dodge, Colorado Springs, April 18, 1881, Film I.
[7] Negotiations between the two parties may be followed in Dodge to Danforth, Colorado Springs, May 8, 1881; Corwin to Danforth, Narrowsburg, New York, June 13, 1881; Danforth to Dodge, Colorado Springs, June 26, 1881; and Dr. Bancroft to Dodge, June 28, 1881, Film I.
[8] *Pueblo Daily Chieftain,* Oct. 28, 1881, Jan. 10, April 23, Nov. 29, 1882.
[9] *Ibid.,* June 11, Nov. 29, 1882. See also "The Medical Department of The Colorado Fuel and Iron Company," p. 106.
[10] Affidavit by D. C. Dodge, undated, Film I.
[11] "The Medical Program of The Colorado Fuel and Iron Company," pp. 106-107.
[12] "Personnel of the Medical Department," pp. 173-174. See also Colorado Fuel and Iron Company, "Predecessor Companies of The Colorado Coal & Iron Co.," p. 208.
[13] In the period from 1880 to 1892, contributions to the hospital fund did exceed expenses. At the time of consolidation, Osgood's agent reported that the coal and iron company had accumulated a surplus of $23,231.56. However, the corporation had used the money only to defray expenses of the medical program and considered the balance a deferred liability. See Glen W. Traer to J. C. Osgood, May 25, 1892.

14 Danforth to miners at Walsenburg, Jan. 31, 1885, Film H.

15 *Ibid.* The vote was recorded on this document by signatures under columns that indicated favor or disfavor.

16 Danforth to miners at El Moro, March 18, 1885, Film H.

17 "The Medical Department of the Colorado Fuel and Iron Company," p. 107. See also Colorado Fuel and Iron Company, "Predecessor Companies of The Colorado Coal & Iron Co.," p. 208.

18 "The Dispensary System." *Camp and Plant,* III (June 27, 1903): 623-624.

19 *Ibid.,* p. 622.

20 *Ibid.,* pp. 621-625; "The Medical Department of The Colorado Fuel and Iron Company," pp. 109-110.

21 "The Medical Department of The Colorado Fuel and Iron Company," pp. 108-109.

22 "Personnel of the Medical Department," pp. 165-170.

23 "The Medical Department of The Colorado Fuel and Iron Company," p. 112; Hennie S. Cottle, "Modern Trained Nursing." *Camp and Plant,* III (March 21, 1903): 247-248.

24 Cottle, "Modern Trained Nursing," pp. 248-249.

25 "Hospital Training for Nurses, Pueblo, Colorado." *Camp and Plant,* III (March 21, 1903): 250-253.

26 "The Medical Department of The Colorado Fuel and Iron Company," p. 113.

27 "Chief Surgeon's 47th Annual Report," pp. 4-5.

28 *Ibid.*; R. W. Corwin, "New Minnequa Hospital of C. F. & I. Co." *Camp and Plant,* II (Aug. 6, 1902): 98; *CF& I Blast,* Dec. 31, 1937.

29 "C. F. & I. Co. Medical Department." *Camp and Plant,* II (Sept. 6, 1902), p. 233.

30 Corwin, "New Minnequa Hospital of C. F. & I. Co.," pp. 99-101.

31 "C. F. & I. Co. Medical Department," pp. 97-123.

32 *Ibid.,* p. 106.

33 *Ibid.,* p. 111.

34 *Ibid.,* pp. 101-102; "Chief Surgeon's 47th Annual Report," p. 5.

35 "Personnel of the Medical Department," p. 170.

36 *Ibid.,* pp. 170-171.

37 *Ibid.,* pp. 171-172.

38 *Ibid.,* pp. 173-174.

39 *Pueblo Daily Chieftain,* Oct. 7, 1883, Aug. 24, 1888: *Camp and Plant,* II (Aug. 6, 1902): 165; Colorado Fuel and Iron Minute Book of Directors, p. 144; Smiley, ed., *Semi-Centennial History of the State of Colorado,* II: 236.

40 Colorado Fuel and Iron Company, *Annual Report of the Medical Department, 1929,* p. 10.

41 *Ibid.,* p. 11; Smiley, ed., *Semi-Centennial History of the State of Colorado,* II: 236.

42 Colorado Fuel and Iron Company, *Annual Report of the Medical Department, 1929,* p. 8. The Colorado Fuel and Iron Corporation transferred the hospital in February 1948 to the Sisters of Charity of Cincinnati, Ohio. A new multistory structure was erected on the site near Lake Minnequa and operates at the present time as the St. Mary-Corwin Hospital.

Chapter 11. Industrial Sociology

1 Colorado Fuel and Iron Company, *Annual Report of the Sociological Department, 1901-1902* (Denver, 1902), pp. 11-12.

2 *Pueblo Daily Chieftain,* Aug. 25, 1901.

3 A report of this investigation is in Colorado Bureau of Labor Statistics, *Biennial Report, 1901-1902,* pp. 149-154.

4 "The Minnequa Works of The Colorado Fuel and Iron Company," p. 154.

[5] It is unclear whether these new employees were natives of Mexico or Americans of Mexican ancestry. See Colorado Coal and Iron Company, *Annual Report, 1879,* p. 6.

[6] U.S. Immigration Commission, *Immigrants in Industries,* Pt. 25, *Japanese and Other Immigrant Races in the Pacific Coast and Rocky Mountain States. Senate Doc.,* No. 633, 61 Cong., 2 Sess. (Washington, 1911), p. 258.

[7] "Camp Engle or Engleville," p. 459.

[8] "Walsen and Robinson Mines," p. 266.

[9] *Pueblo Daily Chieftain,* Oct. 15, 1881, Dec. 7, 1899; Colorado Fuel and Iron Company, *Annual Report of the Sociological Department, 1901-1902,* p. 5.

[10] Colorado Fuel and Iron Company, *Annual Report of the Sociological Department, 1901-1902,* p. 5; George M. Kirk, "Sociological Work Started by Dr. R. W. Corwin Now Embraces Company's Broad Activities." *The Blast,* April 15, 1932.

[11] Colorado Fuel and Iron Company, *Annual Report of the Sociological Department, 1901-1902,* p. 10.

[12] *Ibid.,* p. 3. Information about the activities of the Sociological Department is drawn almost exclusively from the company's publications, namely, annual reports of the department and articles printed in *Camp and Plant.* While the accuracy of these sources may be questioned, there are no other sources for the years from 1901 through 1903 and later.

[13] *Ibid.,* p. 5.

[14] *Ibid.,* p. 16.

[15] *Ibid.*

[16] *Ibid., Annual Report of the Sociological Department, 1903-1904* (Pueblo, 1904), p. 18.

[17] *Ibid.,* pp. 22-23; "Details of the Year's Work of the Sociological Department of The Colorado Fuel and Iron Company." *Camp and Plant,* IV (Aug. 29, 1903): 150.

[18] Colorado Fuel and Iron Company, *Annual Report of the Sociological Department, 1901-1902,* p. 6.

[19] *Ibid., Annual Report of the Sociological Department, 1903-1904,* p. 24.

[20] "Excellent Public Schools." *Camp and Plant,* IV (Dec. 26, 1903): 557-569.

[21] Colorado Fuel and Iron Company, *Annual Report of the Sociological Department, 1902-1903* (1903), p. 14.

[22] *Ibid., Annual Report of the Sociological Department, 1901-1902,* p. 33.

[23] *Ibid.,* p. 14; *Pueblo Daily Chieftain,* Aug. 25, 1901.

[24] Colorado Fuel and Iron Company, *Annual Report of the Sociological Department, 1902-1903,* pp. 15-16; *Annual Report of the Sociological Department, 1903-1904,* pp. 32-33; "Details of the Year's Work of the Sociological Department of The Colorado Fuel and Iron Company," p. 154.

[25] Colorado Fuel and Iron Company, *Annual Report of the Sociological Department, 1903-1904,* p. 34; "The Colorado Polytechnic Association." *Camp and Plant,* IV (Aug. 1, 1903): 60.

[26] Colorado Fuel and Iron Company, *Annual Report of the Sociological Department, 1902-1903,* pp. 16-17; *Annual Report of the Sociological Department, 1903-1903,* p. 34.

[27] *Ibid., Annual Report of the Sociological Department, 1904-1905* (Denver, 1905), p. 24.

[28] *Ibid.,* p. 24.

[29] *Ibid., Annual Report of the Sociological Department, 1901-1902,* p. 36; *Camp and Plant,* pp. 36-37; *Annual Report of the Sociological Department 1904-1905,* p. 28.

[30] Colorado Fuel and Iron Company, *Annual Report of the Sociological Department, 1901-1902,* pp. 36-37; *Annual Report of the Sociological Department, 1904-1905,* p. 28.

31 *Ibid., Annual Report of the Sociological Department, 1902-1903*, pp. 23-25.
32 *Ibid.*
33 *Ibid., Annual Report of the Sociological Department, 1901-1902*, p. 46.
34 *Ibid., Annual Report of the Sociological Department, 1903-1904*, p. 37.
35 "Club Houses and Recreation Halls." *Camp and Plant*, IV (Nov. 21, 1903): 438-439.
36 Colorado Fuel and Iron Company, *Annual Report of the Sociological Department, 1901-1902*, p. 47.
37 *Ibid., Annual Report of the Sociological Department, 1903-1904*, p. 38.
38 *Ibid.*
39 "Houses and Towns: Built by The Colorado Fuel and Iron Company and The Rocky Mountain Coal and Iron Company for Employees." *Camp and Plant*, V (April 9, 1904): 308.
40 *Ibid.*, p. 309.
41 *Ibid.*, p. 312. See also *Camp and Plant*, I (April 5, 1902): 284.
42 Minnequa Town Company, *Annual Report, 1901*, p. 4; *Annual Report, 1902* (Pueblo, 1902), pp. 4-6; *Pueblo Daily Chieftain*, May 24, Aug. 11, 1900; *Camp and Plant*, II (Nov. 29, 1902): 251.
43 Colorado Fuel and Iron Company, *Annual Report of the Sociological Department 1902-1903*, p. 27; "Camp and Plant's First Year." *Camp and Plant*, II (Dec. 13, 1902): 564.
44 Colorado Fuel and Iron Company, *Annual Report of the Sociological Department, 1902-1903*, p. 28.
45 Colorado Fuel and Iron Company, *Annual Report, 1916* (Denver, 1916), pp. 5-6. See also Kirk, "Sociological Work Started by Dr. R. W. Corwin Now Embraces Company's Broad Activities."

Chapter 12. Osgood's Fight with John W. Gates, 1901-1902

1 *Denver Times*, April 4, 5, 1901; *Pueblo Daily Chieftain*, April 5, 6, 1901.
2 *Commercial and Financial Chronicle*, LXXV (Aug. 30, 1902): 458.
3 Jerome Brief P, p. 11.
4 Osgood to Jerome, New York, June 15, 1901, Jerome Collection.
5 Osgood to Jerome, New York, June 22, 1901, *ibid.*
6 Jerome Brief R, p. 5.
7 Affidavits of Osgood and Mitchell in *John J. Mitchell et al. vs. The Colorado Fuel and Iron Company et al.*, Docket 4330, U.S. Circuit Court, District of Colorado, in Federal Records Center, Region 8, Denver.
8 Jerome Brief N, pp. 8-9.
9 *Ibid.*, pp. 2, 12. See also Brief W, pp. 4-5.
10 Osgood to Jerome, New York, April 20, 26, 1901, Jerome Collection.
11 Osgood to Jerome, April 20, 1901, *ibid.*
12 Osgood to Jerome, April 26, 1901, *ibid.*
13 *Ibid.*
14 *Ibid.*
15 Osgood to Jerome, New York, May 5, 1901, *ibid.*
16 Jerome Brief P, p. 1.
17 *Ibid.*, p. 3.
18 Osgood to Jerome, May 10, 1901, Jerome Collection. The correspondence and agreement with the Illinois Trust and Savings Company and with Blair and Company, dated May 10, 1901, is in Colorado Fuel and Iron Company Minute Book of Stockholders, pp. 31-34, 40-69.
19 Jerome Brief P, p. 6.
20 Osgood to Jerome, New York, June 22, 1901, Jerome Collection. See also Brief Q, p. 1.
21 Jerome Brief P, pp. 7-10; Colorado Fuel and Iron Company Minute Book of Stockholders, pp. 23-24.

[22] Jerome Brief Q, p. 2.
[23] *Ibid.*, p. 5; Colorado Fuel and Iron Company Minute Book of Stockholders, p. 77.
[24] "Petition for Writ of Certiorari or Prohibition," p. 2. This document was prepared by the company for the appeal of the U.S. Circuit Court's order in *George F. Bartlett vs. John W. Gates et al.*
[25] *Ibid.*, p. 4.
[26] Transcript of Record: An Appeal from United States Circuit Court District of Colorado, *George F. Bartlett vs. John W. Gates et al., Cross-Complainants vs. George F. Bartlett et al.*, U.S. Circuit Court of Appeals, Eighth District, Docket 4331, p. 13.
[27] Judge's Memorandum, Docket 4330. See also Transcript of Record, Docket 4331.
[28] Transcript of Record, Docket 4331, p. 13.
[29] *Ibid.*, pp. 13-14, 17-18.
[30] Colorado Fuel and Iron Company Minute Book of Directors, p. 269; "Petition for Writ of Certiorari or Prohibition," pp. 8-9.
[31] "Petition for Writ of Certiorari or Prohibition," p. 9.
[32] *Denver Republican*, Aug. 9, 13, 14, 1902.
[33] *Ibid.*, Aug. 14, 1902.
[34] *Ibid.*
[35] *Ibid.*, Aug. 18, 1902.
[36] Jerome Brief S, p. 27.
[37] *Ibid.*, p. 28.
[38] *Ibid.*, p. 24.
[39] Bill of complaint, Docket 4330.
[40] *Ibid.* See also Transcript of Record, Docket 4331, pp. 56-67.
[41] Bill of complaint, Docket 4330.
[42] Order Denying Injunction, *ibid.*
[43] Jerome Brief S, pp. 31-32.
[44] *Ibid.*
[45] Transcript of Record, Docket 4331, pp. 52-65.
[46] *Ibid.*, pp. 70-72.
[47] Cross-bill of complaint, Docket 4331.
[48] Colorado Fuel and Iron Company, Minute Book of Stockholders, p. 78.
[49] Transcript of Record, Docket 4331, pp. 91-156.
[50] *Ibid.*, pp. 295-310, 311-316.
[51] *Ibid.*, pp. 343-344.
[52] *Denver Republican*, Aug. 23, 1902.
[53] *Ibid.*, Aug. 20, 1902.
[54] *Ibid.*, Aug. 25, 1902.
[55] Reprinted from *New York Sunday World* in *ibid.*, Aug. 28, 1902.
[56] Jerome Brief S, p. 34.
[57] *Ibid.*, pp. 37-39; *Denver Republican*, Nov. 12, 1902. Copies of circulars issued by the three committees are in Colorado Fuel and Iron Company Scrapbook.
[58] Ripley to Morawetz, Chicago, Oct. 21, 1902, AT&SF Collection.
[59] *Ibid.*
[60] Extract of letter, Morawetz to Ripley, Oct. 23, 1902, in *ibid.*
[61] Extract of letter, Ripley to Morawetz, Oct. 28, 1902; Ripley to Morawetz, Chicago, Oct. 29, 1902, in *ibid.*
[62] See circulars in Colorado Fuel and Iron Company Scrapbook.
[63] Jerome Brief T, p. 1.
[64] *Denver Republican*, Dec. 10, 1902; *Denver Post*, Dec. 10, 1902.
[65] Colorado Fuel and Iron Company Minute Book of Stockholders, pp. 79-80.
[66] *Ibid.*; *Denver Republican*, Dec. 11, 1902.
[67] Colorado Fuel and Iron Company Minute Book of Directors, p. 294.

68 *Camp and Plant*, II (Dec. 20, 1902): 601-602.
69 *Pueblo Daily Chieftain*, Dec. 11, 12, 1902; *Denver Times*, Dec. 11, 1902.
70 *Denver Times*, Dec. 11, 1902.

Chapter 13. The End of the Osgood Era, 1902-1903

1 Jerome Brief S, p. 2; Colorado Fuel and Iron Company Minute Book of Directors, p. 226.
2 Colorado Fuel and Iron Company, *Annual Report, 1902,* pp. 3-5; *Annual Report, 1903*, p. 4; Minute Book of Directors, p. 249.
3 Colorado Bureau of Labor Statistics, *Biennial Report, 1901-1902,* p. 230. See also *Pueblo Daily Chieftain*, May 15, 24, June 1, 3, 29, Aug. 14, 1902.
4 Jerome Brief S, pp. 9-10.
5 *Ibid.*, p. 12.
6 *Ibid.*, p. 17.
7 *Ibid.*, p. 19; Colorado Fuel and Iron Company Minute Book of Directors, p. 258.
8 Jerome Brief S, pp. 20-22.
9 *Ibid.*, Brief T, pp. 2-3.
10 *Ibid.*, p. 4.
11 Colorado Fuel and Iron Company Minute Book of Executive Committee, pp. 24-25.
12 Jerome Brief T, p. 6.
13 *Ibid.*, p. 7.
14 Colorado Fuel and Iron Company Minute Book of Directors, p. 305; Jerome Brief T, pp. 9-10.
15 Colorado Fuel and Iron Company Minute Book of Directors, pp. 305-306.
16 *Ibid.*, pp. 311-312.
17 *Ibid.*, pp. 131-318.
18 Colorado Fuel and Iron Company Minute Book of Stockholders, pp. 82-83.
19 *Engineering and Mining Journal*, LXXVI (Aug. 29, 1903): 324-325.
20 Colorado Fuel and Iron Company Minute Book of Directors, pp. 325-326; *Denver Republican*, July 26, 1903; "The New President and Chairman." *Camp and Plant*, IV (Oct. 3, 1903): 283.
21 *Denver Republican*, Aug. 20, 1903.
22 "The Late A. C. Cass," p. 186; "Julian Abbot Kebler," p. 353; "The Late John Lathrop Jerome." *Camp and Plant*, V (April 30, 1904): 398.
23 Colorado Fuel and Iron Company Minute Book of Executive Committee, p. 41. The manner in which Osgood acquired the Victor Coal and Coke Company and reorganized it as the Victor Fuel Company is discussed in Jerome Brief J, pp. 16-21.
24 Colorado State Inspector of Coal Mines, *Biennial Report, 1903-1904*, pp. 96-97, 150, 156, 160.
25 Julian Kennedy to Messrs. Blair and Company, Pittsburgh, Aug. 6, 1901, Film E.
26 Ripley to Morawetz, Chicago, Aug. 9, 1904, in AT&SF Collection.
27 *Ibid.* See also Ripley to Morawetz, Chicago, Aug. 19, 1904.
28 Ripley to Morawetz, Aug. 9, 1904.
29 *Ibid.* See also Colorado State Inspector of Coal Mines, *Biennial Report, 1903-1904*, pp. 125-126.
30 Ripley to Morawetz, Aug. 9, 1904.
31 Colorado State Inspector of Coal Mines, *Biennial Report, 1903-1904*, pp. 150, 156, 160.
32 *Commercial and Financial Chronicle*, LXXXIX (Aug. 14, 1909): 415; LXXXX (June 25, 1910): 1683.
33 Vaughn Mechau, "Redstone on the Crystal." *Denver Westerner's Brand Book*, IV (1948): 52.

[34] *Ibid.* The largest single stockholder in the Victor-American Fuel Company at the time of his death, Osgood also held large interests in the marble deposits of Gunnison County, in addition to real estate in eastern cities. Although he had previously transferred much of his property to his third wife, Lucille Reed Osgood, whom he married in 1920, she nonetheless inherited over four million dollars from his estate. See Duane Vandenbusche and Rex Myers, *Marble, Colorado: City of Stone* (Denver, 1970), pp. 36-42; *Denver Post*, July 14, 1932.

Epilogue

[1] Colorado State Planning Division, *Colorado Year Book, 1962-1964* (Denver, 1965), p. 550.
[2] Colorado Fuel and Iron Corporation, "History of Coal Mines and Coke Plants," pp. 3, 21, 24. Typescript in Mining Department, Pueblo.
[3] *The Blast*, Aug. 15, 1932.
[4] H. B. Carpenter, "By-Product Coke Plant at Minnequa," *Colorado Fuel & Iron Company Industrial Bulletin*, III (Jan. 31, 1918): 6-8, 14.
[5] C. C. Crawford, "CF&I Steel at Pueblo," *Iron and Steel Engineer*, XXXIX (May 1962): 5-6.
[6] Colorado State Inspector of Coal Mines, *Annual Report, 1920* (Denver, 1921), *passim*; Colorado Fuel and Iron Corporation, "History of Coal Mines and Coke Plants," pp. 1-25.
[7] Colorado Fuel and Iron Corporation, "History of Coal Mines and Coke Plants," pp. 7-18.
[8] *Ibid.*, pp. 19-21; *Pueblo Daily Chieftain*, Aug. 10, 1901.
[9] Colorado Fuel and Iron Corporation, "History of Coal Mines and Coke Plants," pp. 22-25.
[10] Colorado Fuel and Iron Corporation, "Statement of Coal Mined by Predecessor Companies prior to December 31, 1965," in Mining Department, Pueblo.
[11] Colorado State Inspector of Coal Mines, *Annual Report, 1940* (Denver, 1941), p. 14. The Pictou Mine was idle in 1940.
[12] *CF&I Blast*, April 23, June 4, 1956; Colorado Fuel and Iron Corporation, "Statement of Coal Mined by Predecessor Companies prior to December 31, 1965."
[13] Colorado Fuel and Iron Company, "Raw Materials for Colorado's Steel Industry," p. 4. Typescript in Mining Department, Pueblo.
[14] "The Minnequa Works of The Colorado Fuel and Iron Company," pp. 149-154; Colorado Fuel and Iron Company, *Annual Report, 1903*, p. 4.
[15] Colorado Fuel and Iron Company, *Annual Report, 1905* (Denver, 1905), p. 3.
[16] *Ibid.*, *Annual Report, 1906* (Denver, 1906), pp. 3-4; *Annual Report, 1907* (Denver, 1907), p. 4.
[17] *Ibid.*, *Annual Report, 1913* (Denver, 1913), p. 3; *Annual Report, 1916* (Denver, 1916), p. 3.
[18] *Ibid.*, *Annual Report, 1914* (Denver, 1914), pp. 3-4.
[19] *Coal Age*, IV (Sept. 27, 1913), p. 456, (Nov. 29, 1913), pp. 816-817.
[20] Colorado Fuel and Iron Company, *Annual Report, 1914*, pp. 4-5.
[21] *Ibid.*, p. 4.
[22] *Ibid.*, p. 5.
[23] *Ibid.*, pp. 5-6.
[24] *Ibid.*, p. 7.
[25] *Ibid.*, p. 8.
[26] *Ibid.*, p. 9.
[27] *Ibid.*, p. 10.
[28] *Ibid.*

29 U.S. Senate, *Industrial Relations: Final Report and Testimony Submitted to the Congress by the Commission on Industrial Relations Created by the Act of August 23, 1912. Senate Doc.* No. 415, 64 Cong., 1 Sess. (11 vols., Washington, 1916). Testimony related to the strike in Colorado is in vols. 7-9. See also Graham Adams, Jr., *Age of Industrial Violence, 1910-1915: The Activities and Findings of the United States Commission on Industrial Relations* (New York, 1966), Chap. 7.

30 See Barron B. Beshoar, *Out of the Depths:The Story of John R. Lawson, a Labor Leader* (Denver 1943), *passim*; and George S. McGovern and Leonard F. Guttridge, *The Great Coalfield War* (Boston, 1972), *passim.*

31 See Irving Bernstein, *The Lean Years: The Story of the American Worker, 1920-1933* (Boston, 1960), pp. 157-164; David Brophy, *Steelworkers in America:The Nonunion Era* (Cambridge, Massachusetts, 1960), p. 226. The origin and early application of the representation plans are discussed critically in Ben M. Selekman, *Employees' Representation in Steel Works* (New York, 1924) and Selekman and Mary Van Kleek, *Employees' Representation in Coal Mines* (New York, 1924).

32 The disappearance of a separate annual report for the Sociological Department after 1908 and the combination of that publication with the Medical Department's annual statement of activities suggests that social improvement and education had been given lesser priorities. The author has found no record of annual expenditures for the Sociological Department after 1903.

33 Gaddis' testimony may be found in U.S. Senate, Industrial Relations: Final Report and Testimony Submitted to the Congress by the Commission on Industrial Relations, IX: 8483-8531. Specific criticisms mentioned here may be found on pp. 8491-8503.

34 See, for example, testimony of Rev. Daniel Spencer McCorkle, minister of the Presbyterian Church in Sunrise, Wyoming, a community that was not directly affected by the strike of 1913-1914. *Ibid.*, IX: 8532-8592.

35 Selekman, *Employees' Representation in Steel Works*, pp. 27-28.

36 "Industrial Democracy Established by Colorado Fuel and Iron Co.," *Engineering and Mining Journal*, CIV (Oct. 27, 1917): 748.

37 Bernstein, *The Lean Years*, p. 164.

38 "C.I.O. vs. C. F. & I.," *Business Week*, Aug. 15, 1942, pp. 92-93.

39 Colorado Fuel and Iron Company, *Annual Report, 1916*, p. 3; *Annual Report, 1917* (Denver, 1918), p. 13. The company changed its fiscal year in 1917. Of the two annual reports issued that year, the one cited is for the calendar year.

40 *Ibid., Annual Report, 1920* (Denver, 1921), p. 3; *Annual Report, 1921* (Denver, 1922), p. 3.

41 *Ibid., Annual Report, 1924* (Denver, 1925), p. 4.

42 *Ibid.*, "Plan of Reorganization, March 1, 1935," pp. 1-2; *CF&I Blast*, June 26, 1936; "A New Day for Bankrupt Steel Mill." *News Week*, VIII (July 4, 1936): 31-32.

43 "A New Day for Bankrupt Steel Mill," p. 32.

44 *Commercial and Financial Chronicle*, CLI (Sept. 14, 1940): 1567.

45 *Ibid.*, CXLIV (June 5, 1937): 3833, (June 26, 1937): 4339., CXLV (Oct. 9, 1937): 2386; *CF&I Blast*, June 4, 1937. The corporation stated at this time that the California Wire Cloth Co. was its first manufacturing facility located outside the state of Colorado. This was incorrect. The Pueblo firm had organized in 1902 the wholly owned subsidiary, the Minnequa Cooperage Company, to erect and operate near Little Rock, Arkansas, a plant to manufacture staves and headings for kegs used in the nail, spike, and bolt mills. See Minnequa Cooperage Company Minute Book, p. 1; Edward Price, "Making Staves and Heading." *Camp and Plant*, IV (Dec. 5, 1903): 485-487.

[46] *CF&I Blast*, June 4, 1937.

[47] *Ibid.*, Jan.-Feb., 1972; Hafen, ed., *Colorado and Its People*, I: 620.

[48] Colorado Fuel and Iron Corporation, *Trail-Blazing: Accomplishments of Men Working Together* (1949), pp. 18-19.

[49] *Ibid.*, p. 19. See also Colorado Fuel and Iron Corporation, "75th Anniversary Dinner," (1947), p. 6.

[50] Colorado Fuel and Iron Corporation, "Raw Materials for Colorado's Steel Industry," p. 6. See also *CF&I Blast*, June 1971.

[51] C. Langdon White, *Is the West Making the Grade in the Steel Industry?* (Stanford, California, 1956), pp. 7-16; "Steel in the West." *Fortune*, XXXI (Feb. 1945): 30-33, 258-261.

[52] "Merger in Steel."*Business Week*, July 7, 1945, p. 58.

[53] See the following articles in *ibid.*: "Steel for the West—and Kaiser?" July 21, 1945, pp. 15-16; "Steel Bows Out," Aug. 18, 1945, pp. 40, 42; "Bid for Geneva," Sept. 22, 1945, p. 30.

[54] *New York Times*, May 2, July 13, 1946.

[55] *Ibid.*, Dec. 24, 1944; "Chapter Ends." *Business Week*, Jan. 6, 1945, pp. 20-21.

[56] "Merger in Steel," pp. 56, 58; *Commercial and Financial Chronicle*, CLXII (Oct. 15, 1945): 1765.

[57] Colorado Fuel and Iron Corporation, *Trail Blazing*, pp. 20, 30.

[58] *CF&I Blast*, Nov. 21, 1952; "Two-Way Stretch for Steel Company." *Business Week*, June 20, 1953, p. 190.

[59] "Two-Way Stretch for Steel Company," p. 190; *New York Times*, Jan. 16, 1952.

[60] "Two-Way Stretch for Steel Company," p. 190; *New York Times*, Dec. 9, 1952.

[61] "Two-Way Stretch for Steel Company," p. 188; Colorado Fuel and Iron Corporation, *Trail Blazing*, pp. 22-24; *CF&I Blast*, Jan. 28, 1949.

[62] "A New Spur for Steel's 'Also Ran'." *Business Week*, May 13, 1967, pp. 188-190, 192; *CF&I Steel Blast*, Dec. 1970. See also *New York Times*, Sept. 15, Oct. 3, 1960, March 14, 1962.

[63] *CF&I Steel Blast*, Jan.-March, 1973.

[64] Colorado Fuel and Iron Corporation, "History of The Colorado Fuel and Iron Corporation," pp. 9-10. Typescript of publicity release in Western History Collections, University of Colorado.

[65] *CF&I Steel Blast*, Oct.-Nov., 1972, April-June, 1973, Jan.-March, 1974.

[66] *Ibid.*, Nov.-Dec., 1968, May, 1969, March-May, 1971.

[67] "A New Spur for Steel's 'Also Ran'." pp. 189-190; *CF&I Steel Blast*, May 15, June 24, 1967.

[68] *Ibid.*, May 1969.

[69] *Ibid.* The Rocky Mountain Paving Company of Colorado Springs was acquired in 1972 and renamed "Fountain Sand and Gravel Company." The Pueblo firm also purchased in late 1973 and early 1974 the Pueblo Compressed Steel Corporation and Denver Metals Company (formerly Newell Salvage), shredders of junk cars, the end product being recycled in steel furnaces. See *CF&I Steel Blast*, Sept-Oct., 1972, Jan.-March, 1974; *Denver Post*, March 21, 1974.

[70] *CF&I Steel Blast*, July 1969.

[71] *Ibid.*, Jan.-Feb. 1972.

[72] *Ibid.*, April-June 1972.

[73] See advertisement inserted in *ibid.*, Feb. 1969.

Bibliographical Essay

The principal collection of historical data for reconstructing the origin and evolution of the Colorado Fuel and Iron Company and predecessor enterprises in the years from 1872 to 1903 is located in the administrative offices and operating departments of what is now the CF&I Steel Corporation, Pueblo, Colorado. These materials, for the most part, are in the office of the secretary, the publicity, personnel, mining, and engineering departments, and land and water department. They consist of correspondence; contracts and agreements; employment records for mills and mines; reports pertaining to manufacturing, mining, marketing, land and water acquisitions; geological exploration reports; medical and sociological programs; journals and ledgers; patents, deeds, and abstracts; engineering and surveyor's plats and maps; minutes of meetings of directors, executive committees, and stockholders; and annual reports to stockholders and bondholders. The several predecessor enterprises for which records may be found in the CF&I Steel Corporation are as follows: Aspen and Western Railway Co., Central Colorado Improvement Co., Colorado Coal and Iron Co., Colorado Coal and Iron Development Co., Colorado Coal and Steel Works Co., Colorado Fuel Co., Colorado Fuel and Iron Co., Colorado-Grand Development Co., Colorado Supply Co., Colorado and Utah Railway Co., Colorado and Wyoming Railway Co., Crystal River Land and Improvement Co., Crystal River Railroad Co., Crystal River Railway Co., Crystal River Toll Road Co., Denver Fuel Co., Elk Mountain Fuel Co., Grand River Coal and Coke Co., Hartville Iron Co., Huerfano Land Association, Las Animas and Western Railway Co., Minnequa Cooperage Co., Minnequa Land and Water Co., Minnequa Town Co., Mountain Telegraph Co., Pueblo Realty Trust Co., Redstone Improvement Co., Rocky Mountain Coal and Iron Co., Rocky Mountain Timber Co., Southern Colorado Coal Co., Southern Colorado Coal and Town Co., Southern Land Co., Trinidad Fuel Co., Wyoming Railway and Iron Co., and Yampa River Coal Co.

The Colorado Fuel and Iron Co. prepared about 1923 a history, largely statistical in character, of its operations, with emphasis on the activities of predecessor enterprises. This 214-page typescript, entitled "Predecessor Companies of The Colorado Coal & Iron Co.: The Central Colorado Improvement Co., The Southern Colorado Coal & Town Co., The Colorado Coal & Steel Works Co., and The Colorado Coal and Iron Co.," contains data taken principally from financial and production records in the Comptroller's Office.

The CF&I Steel Corporation and its predecessors have published newspapers and magazines that are valuable sources of information about business activities in the years from 1872 to 1903, and later. The publications are *Camp and Plant*, 5 vols., Dec. 14, 1901-April 30, 1904, issued weekly and monthly; *The Colorado Fuel & Iron Company Industrial Bulletin*, 14 vols., Nov. 1915-Feb. 1929, issued monthly; *The Steel Works Blast*, Feb. 2, 1923-July 8, 1929, issued weekly; *The Blast*, Aug. 2, 1929-March 31, 1933, issued weekly; and *CF&I Blast* (*CF&I Steel Blast*), March 3, 1936 to date, issued weekly, monthly, bi-monthly and quarterly. Articles, reports, and obituaries in these publications contain a rich store of information about the CF&I Steel Corporation's predecessors in the nineteenth and twentieth centuries.

In addition to offical publications of the Colorado Fuel and Iron Co., its predecessors and successors, periodicals and newspapers are an indispensible source of information about Colorado's iron and steel, as well as the coal-mining industry. Particularly useful for this study are *Coal Age*, 30 vols., Jan. 1916-May 1934, issued weekly and monthly; *The Engineering and Mining Journal*, 81 vols., Jan. 1874-Dec. 1917, issued weekly; and *The Commercial and Financial Chronicle*, 162 vols., July 6, 1878-Oct. 15, 1945, issued weekly. Newspapers published in Denver, particularly the *Denver Post*, 1895-1975, daily; the *Denver Republican*, 1887-1903, daily; the *Denver Tribune-Republican*, 1884-1886, weekly; *Denver Tribune*, 1871-1884, daily; and the *Rocky Mountain News*, 1872-1975, daily, afford broad coverage of events pertaining to the Colorado Fuel and Iron Co., its predecessors and successors. Equally useful are Pueblo newspapers, especially the *Colorado Chieftain*, 1868-1873, weekly; and the *Pueblo Daily Chieftain*, 1873-1903. Fuel operations in Fremont and Las Animas counties, where the Pueblo company's principal coal mines were located, require careful reading of the *Canon City Times*, 1872-1877, weekly; the *Trinidad Daily Advertiser*, 1883-1888; and *Trinidad Chronicle-News*, 1899-1903, weekly.

Numerous manuscript collections provide valuable insights into people and events related to the Pueblo heavy industry's far-ranging business activities. Housed in the Library of The State Historical Society of Colorado are the papers of General William J. Palmer, William A. Bell, and Robert F. Weitbrec, all founders or early participants in the Central Colorado Improvement Co., the Southern Colorado Coal and Town Co., and the Colorado Coal and Iron Co. The John L. Jerome papers reveal the ventures of the so-called "Iowa Group" in the Denver Fuel Co., Colorado Fuel Co., Elk Mountain Fuel Co., and other enterprises organized by John C. Osgood and close associates. Since Jerome was treasurer of the Colorado Fuel and Iron Co. from 1893 to 1903, his papers offer an "insider's" view of that corporation in its first ten years of operations. Also in The State Historical Society of Colorado is the Denver and Rio Grande Archives, a collection of correspondence and documents pertaining to the Denver and Rio Grande Railway (1870-1885) and its successor, the Denver and Rio Grande Railroad (1885-1935). These materials illuminate the origin and development of the railroad and its relations with the Colorado Coal and Iron Co. and with predecessor enterprises.

Other depositories contain manuscripts that relate to the Colorado Fuel and Iron Co., predecessors, and successors. The Charles Leaming Tutt Library of Colorado College, Colorado Springs, possesses the Denver and Rio Grande and the Central Colorado Improvement Company scrapbooks, two volumes which include correspondence and documents from the early years of the respective enterprises. The historical records of the Atchison, Topeka and Santa Fe Railway Co., including records of subsidiary coal firms in Colorado, are in The Kansas State Historical Society, Topeka. The papers of William Blackmore, a partner of General Palmer in early land speculations and business ventures, are in The New Mexico Records Center and Archives, and The New Mexico Historical Society, Santa Fe.

A variety of historical records pertinent to the Pueblo enterprise may be found in the National Archives and Records Service depositories. Correspondence of the U.S. Attorney General, the U.S. Attorney for Colorado, officials of the Department of the Interior, including the General Land Office, related to land fraud in Colorado, may be found in Record Groups 21 and 49, National Archives, Washington, D.C. Records of the U.S. Circuit Court for the District of Colorado are located in the Branch Archives, Region 8, NARS, Denver Federal Center. These include several suits initiated by the Department of Justice against alleged perpetrators of fraud in securing coal lands from the public domain. Legal actions were brought against the Southern Colorado Coal and Town Co., the Colorado Coal and Iron Co., the Trinidad Coal and Coking Co., and the Gray Creek Coal and Coking Co., all operating in Las Animas County, as well as numerous other coal-mining enterprises elsewhere in the state. Other federal court cases particularly pertinent to the history of the Colorado Fuel and Iron Co. and predecessors are *Edmund D. Davis vs. The Colorado Coal and Iron Co.*, Docket 1993 (1886); *Henry K., Pomeroy and Cornelius Gold, Trustees, vs. The Colorado Coal and Iron Development Co.*, Dockets 3491 and 3492 (1896); *John J. Mitchell, James A. Blair, John W. Gates and Arthur J. Singer vs. The Colorado Fuel and Iron Co., John C. Osgood, Julian A. Kebler, Alfred C. Cass, Dennis Sullivan, and David C. Beaman*, Docket 4330 (1902); *George F. Bartlett vs. John W. Gates, John J. Mitchell, James A. Blair, and John Lambert,* Docket 4331 (1902).

Pueblo's industrial giant has been the subject of many unpublished scholarly investigations. The first comprehensive examination of the iron and steel manufacturer's evolution was Lois Marguerite Gaynor's "History of the Colorado Fuel and Iron Company and Constituent Companies, 1872-1933," M.A. thesis, University of Colorado, 1936. The following works treat important aspects of the activities of the Colorado Fuel and Iron Co. and its predecessors: Paul Stewart Logan's "The History of the Denver and Rio Grande Railway, 1871-1888," M.A. thesis, University of Colorado, 1931; Robert George Huddleston's "The El Moro Coal Lands: A Study of Fraudulent Preemption of Mineral Lands in Colorado in the 1870's and Their Relation to the Development of the Denver and Rio Grande Railroad," M.A. thesis, The Ohio State University, 1968; and Howard K. Wilson's "A Study of Paternalism in The Colorado Fuel and Iron Company Under John C. Osgood, 1892-1903," M.A. thesis, University of Denver, 1967. Labor-management relations of the Pueblo firm is a theme that runs throughout Donald Joseph McClurg's "Labor Organization in the Coal Mines of Colorado, 1878-1933," Ph.D. dissertation, University of California, 1959. The coal miners' strike of 1884-1885, in which the Colorado Coal and Iron Co. played a crucial role, is treated at some length by Gene Ronald Marlatt's "Joseph R. Buchanan: Spokesman for Labor During the Populist and Progressive Eras," Ph.D. dissertation, University of Colorado, 1975.

Public documents are indispensible sources for a study of the Colorado Fuel and Iron Co. State publications of particular value are the Colorado Bureau of Labor Statistics, *Reports, 1887-1904,* 8 vols., and Colorado State Inspector of Coal Mines, *Reports, 1884-1974,* 77 vols. Federal publications include James M. Swank, "Iron: Iron Ore and Its Products," in *Mineral Resources of the United States,* edited by Albert Williams, Jr. *House Misc. Doc.*, No. 75. 48 Cong., 1 Sess. (Washington, 1883). A record of labor disputes appears in U.S. Commissioner of Labor, *Report on Strikes and Lockouts, Jan. 1, 1881-Dec. 31, 1886. House Ex. Doc.*, No. 1, Pt. 5, Vol. 5. 50 Cong., 1 Sess. (5 vols., Washington, 1887). This should be supplemented with first-hand reports of the strike of 1913-1914 in U.S. Senate, Industrial Relations: *Final Report and Testimony Submitted to the Congress by the Commission on Industrial Relations Created by the Act of August 23, 1912. Senate Doc.*, No. 415. 64 Cong., 1 Sess. (11 vols., Washington, 1916). Testimony pertaining to the Colorado miners' strike appears in volumes seven through nine. The evolution of the iron and steel industry and the growth of metallurgical combinations are discussed at length in U.S. Commission of Corporations, *Report on the Steel Industry* (Washington, 1911).

The evolution of iron and steel manufacturing in America has been the subject of numerous books. Especially valuable for placing the Colorado Fuel and Iron Co. in its industrial context are Arthur Cecil Bining, *The Pennsylvania Iron Manufacture in the Eighteenth Century* (Harrisburg: Pennsylvania Historical Commission, 1938); Herbert N. Casson, *The Romance of Steel: The Story of a Thousand Millionaires* (New York: A.S. Barnes & Company, 1907); Victory S. Clark, *History of Manufactures in the United States* (3 vols., New York: McGraw-Hill Book Company, Inc., 1929); Burton J. Hendrick, *The Life of Andrew Carnegie* (2 vols., New York: Doubleday, Doran & Company, Inc., 1932); Gertrude G. Schroeder, *The Growth of Major Steel Companies, 1900-1950* (Baltimore: The

Johns Hopkins Press, 1953); James M. Swank, *History of the Manufacture of Iron in All Ages, and Particularly in the United States from Colonial Times to 1891* (2nd ed., Philadelphia, 1892; New York: Burt Franklin, 1969); Peter Temin, *Iron and Steel in Nineteenth-Century America: An Economic Inquiry* (Cambridge, Massachusetts: The M. I. T. Press, 1964); Joseph Frazier Wall, *Andrew Carnegie* (New York: Oxford University Press, 1970); and Lloyd Wendt and Herman Kogan, *Bet a Million! The Story of John W. Gates* (Indianapolis: The Bobbs-Merrill Company, Publishers, 1948).

The state and regional setting for the growth of Colorado's leading heavy industry is readily available in numerous books. The most useful of these are James H. Baker and LeRoy R. Hafen, eds., *History of Colorado* (5 vols., Denver: Linderman Co., Inc., 1927); Harold Hathaway Dunham, *Government Handout: The Study of the Administration of the Public Lands, 1875-1891* (New York: Da Capo Press, c. 1941, 1971); LeRoy R. Hafen, ed., *Colorado and Its People: A Narrative and Topical History of the Centennial State* (4 vols., New York: Lewis Historical Publishing Co., Inc., 1948); Frank Hall, *History of the State of Colorado* (4 vols., Chicago: The Blakely Printing Company, 1889-1895); *History of the Arkansas Valley, Colorado* (Chicago: O. L. Baskin & Co., Publishers, 1880); *History of the City of Denver, Arapahoe County, and Colorado* (Chicago: O. L. Baskin & Co., Publishers, 1880); Jerome C. Smiley, ed., *Semi-Centennial History of the State of Colorado* (2 vols., Chicago and New York: The Lewis Publishing Company, 1913); Marshall Sprague, *Newport of the Rockies: The Life and Good Times of Colorado Springs* (Denver: Sage Books, 1961); Wilbur Fisk Stone, *History of Colorado* (4 vols., Chicago: The S. J. Clarke Publishing Co., 1918-1919); Carl Ubbelohde, Maxine Benson, and Duane Smith, *A Colorado History* (3rd ed., Boulder, Colorado: Pruett Publishing Co., 1972); Milo Lee Whittaker, *Pathfinders and Pioneers of the Pueblo Region: Comprising a History of Pueblo from the Earliest Times* (Pueblo: The Franklin Press, 1917); and James F. Willard, ed., *The Union Colony at Greeley, Colorado, 1869-1871* (Boulder: University of Colorado, 1918).

Railroads played conspicuous roles in the evolution of the Colorado Fuel and Iron Co., in part because the Denver and Rio Grande Railway was the parent enterprise, and also because the iron and steel producer, while dependent on that and other carriers for moving bulk materials in the form of ore, coal, and limestone, as well as finished products, came into being primarily to supply railroads with tracks and accessories. Studies pertinent to the development of Pueblo's leading industry are George L. Anderson, *General William J. Palmer: A Decade of Colorado Railroad Building, 1870-1880.* Colorado College Publication No. 208 (Colorado Springs, 1936) and *Kansas West* (San Marino, California: Golden West Books, 1963); Robert G. Athearn, *Rebel of the Rockies: A History of the Denver and Rio Grande Western Railroad* (New Haven and London: Yale University Press, 1962); William A. Bell, *New Tracts in North America: A Journal of Travel and Adventure Whilst Engaged in the Survey of a Southern Railroad to the Pacific During 1867-1868* (2nd ed., London: Chapman and Hall; New York: Scribner, Welfore & Co., 1870); Herbert O. Brayer, *A Case Study in Economic Development of the West* (2 vols., Denver: Bradford-Robinson, 1949); John S. Fisher, *A Builder of the West: The Life of William Jackson Palmer* (Caldwell, Idaho: The Caxton Printers, Ltd., 1939); Irving Howbert, *Memories of a Lifetime in the Pike's Peak Region* (New York and London: G. P. Putnam's Sons, 1925); John J. Lipsey, *The Lives of James John Hagerman: Builder of The Colorado Midland Railway.* Edited by Edith Powell (Denver, Colorado: Golden Bell Press, 1968); David F. Myrick, *New Mexico's Railroads: An Historical Survey* (Golden, Colorado: Colorado Railroad Museum, 1970); Richard C. Overton, *Gulf to the Rockies: The Heritage of the Fort Worth and Denver—Colorado and Southern Railways* (Austin: University of Texas Press, 1953); and M. C. Poor, *Denver South Park & Pacific: A History of the Denver South Park & Pacific and Allied Narrow Gauge Lines of the Colorado & Southern Railway Company* (Denver: Rocky Mountain Railroad Club, 1949).

Other books portray important aspects of the Colorado Fuel and Iron Company's history. Among these, the most important are Graham Adams, Jr., *Age of Industrial Violence 1910-1915: The Activities and Findings of the United States Commission on Industrial Relations* (New York and London: Columbia University Press, 1966); Irving Bernstein, *The Lean Years: A History of the American Worker, 1920-1933* (Boston: Houghton Mifflin Company, 1966) and *Turbulent Years: A History of the American Worker, 1933-1941* (Boston: Houghton Mifflin Company, 1969); Howard N. Eavenson, *The First Century and a Quarter of the American Coal Industry* (Pittsburgh, Pa.: Privately printed, 1942); Raymond B. Fosdick, *John D. Rockefeller, Jr.: A Portrait* (New York: Harper & Row, Publishers, 1956); Jim Berry Pearson, *The Maxwell Land Grant* (Norman: University of Oklahoma Press, 1961); Ben M. Selekman, *Employees' Representation in Steel Works: A Study of the Industrial Representation Plan of the Minnequa Steel Works of the Colorado Fuel and Iron Company* (New York: Russell Sage Foundation, 1924); Ben M. Selekman and Mary Van Kleeck, *Employees' Representation in Coal Mines:A Study of the Industrial Representation Plan of The Colorado Fuel and Iron Company* (New York: Russell Sage Foundation, 1924); and Langdon C. White, *Is the West Making the Grade in the Steel Industry?* (Stanford, California: Graduate School of Business, Stanford University, 1956).

Index